JOURNAL OF CYBER SECURITY AND MOBILITY

Volume 4, No. 1 (January 2015)

Special Issue on
Resilient and Trustworthy IoT Systems

Guest Editor:
Geir M. Køien

JOURNAL OF CYBER SECURITY AND MOBILITY

Editors-in-Chief
Ashutosh Dutta, AT&T, USA
Ruby Lee, Princeton University, USA
Neeli R. Prasad, CTIF-USA, Aalborg University, Denmark

Associate Editor
Shweta Jain, York College CUNY, USA

Steering Board
H. Vincent Poor, Princeton University, USA
Ramjee Prasad, CTIF, Aalborg University, Denmark
Parag Pruthi, NIKSUN, USA

Advisors
R. Chandramouli, Stevens Institute of Technology, USA
Anand R. Prasad, NEC, Japan
Frank Reichert, Faculty of Engineering & Science University of Agder, Norway
Vimal Solanki, Corporate Strategy & Intel Office, McAfee, Inc, USA

Editorial Board

Sateesh Addepalli, CISCO Systems, USA
Mahbubul Alam, CISCO Systems, USA
Jiang Bian, University of Arkansas for Medical Sciences, USA
Tsunehiko Chiba, Nokia Siemens Networks, Japan
Debabrata Das, IIIT Bangalore, India
Subir Das, Telcordia ATS, USA
Tassos Dimitriou, Athens Institute of Technology, Greece
Pramod Jamkhedkar, Princeton, USA
Eduard Jorswieck, Dresden University of Technology, Germany
LingFei Lai, University of Arkansas at Little Rock, USA
Yingbin Liang, Syracuse University, USA
Fuchun J. Lin, Telcordia, USA
Rafa Marin Lopez, University of Murcia, Spain
Seshadri Mohan, University of Arkansas at Little

Rock, USA
Rasmus Hjorth Nielsen, Aalborg University, Denmark
Yoshihiro Ohba, Toshiba, Japan
Rajarshi Sanyal, Belgacom, Belgium
Andreas U. Schmidt, Novalyst, Germany
Remzi Seker, University of Arkansas at Little Rock, USA
K.P. Subbalakshmi, Stevens Institute of Technology, USA
Reza Tadayoni, Aalborg University, Denmark
Wei Wei, Xi'an University of Technology, China
Hidetoshi Yokota, KDDI Labs, USA
Geir M. Køien, University of Agder, Norway
Nayak Debu, Information and Wireless Security IIT Bombay

Aim

Journal of Cyber Security and Mobility provides an in-depth and holistic view of security and solutions from practical to theoretical aspects. It covers topics that are equally valuable for practitioners as well as those new in the field.

Scope

The journal covers security issues in cyber space and solutions thereof. As cyber space has moved towards the wireless/mobile world, issues in wireless/mobile communications will also be published. The publication will take a holistic view. Some example topics are: security in mobile networks, security and mobility optimization, cyber security, cloud security, Internet of Things (IoT) and machine-to-machine technologies.

Published, sold and distributed by:
River Publishers
Niels Jernes Vej 10
9220 Aalborg Ø
Denmark

Tel.: +45369953197
www.riverpublishers.com

Journal of Cyber Security and Mobility is published four times a year.
Publication programme, 2015: Volume 4 (4 issues)

ISSN 2245-1439 (Print Version)
ISSN 2245-4578 (Online Version)
ISBN 978-87-93237-66-7 (this issue)

JOURNAL OF CYBER SECURITY AND MOBILITY COMMUNICATIONS

Volume 4, No. 1 (January 2015)

Editorial Foreword:
Special Issue on Resilient and Trustworthy IoT Systems

The internet of things (IoT) is by now nothing new, but widespread adoption and increasing dependence on IoT services mean that we must ensure that the IoT systems we design, develop, and deploy are resilient and trustworthy. Almost universal availability is expected, yet individual devices will routinely fail and/or be compromised. Despite this, the services should be resilient and trustworthy. The environments where the devices will be deployed will range from protected and controlled environments to potentially very hostile environments where the exposure is extreme.

The cyber security landscape has changed drastically over the last decade. We have adversaries engaging in cyber warfare, organized crime, industrial espionage, petty/opportunistic theft, and privacy invasions. Privacy has become more important, which is by many seen as a prerequisite for human trust in IoT systems. For the IoT systems to remain trustworthy, they need to have credible defenses and be able to detect and respond to incidents.

We have four contributions to this special issue. The first contribution, "Torrent-based Dissemination in Infrastructure-less Wireless Networks" by Kyriakos Manousakis et al, has its roots in peer-to-peer mobile ad hoc networks. This contribution highlights robustness in content dissemination in peer-to-peer mobile ad hoc networks. These networks are subject to disruptions due to erratic link performance and intermittent connectivity. The approach used, called SISTO, is a fully distributed and torrent-based solution. The authors highlight four main features: 1) freedom from reliance on infrastructure; 2) network and topology aware selection of information sources; 3) robust multiple-path routing of content via a proactive peer selection technique; and 4) an integrated distributed content discovery capability. There exist a wide set of network scenarios where these capabilities are useful, such as first responder and disaster recovery situations, and military and tactical operations, which require applications and protocols to function in a purely ad hoc peer-to-peer fashion. SISTO allows IoT devices to provide content in a robust and reliable manner.

The second contribution, titled "Cyber Security for Intelligent World with Internet of Things and Machine to Machine Communication" by Vandana Rohokale and Ramjee Prasad, addresses security for IoT and machine-to-machine (m2m) directly. The paper highlights the diverse and heterogeneous reality of IoT and m2m. The devices can be wired or wireless, they can range from simple RFID tags to relatively powerful 32-bit devices, and they are deployed in a multitude of different locations. Some will be very exposed to hacking, while others will enjoy a relatively safe and secure environment. The devices are subject to both local attacks, possibly with physical intrusion, and global attacks over the networks. The authors investigate state of the art in security provisions for IoT and m2m communications. The authors emphasize security solutions that can grow together with the systems, and they recognize that this is going to be a continuous process. They see role-based access control (RBAC) mechanisms as playing a vital role in the robustness of the cyber security solution development for these services. They also have high hopes for trust-level-based authentication mechanisms and speculate that it may be utilized to provide robust and secure communication.

The third contribution is an effort in improving privacy for an IoT context. The paper, "How to use garbling for privacy preserving electronic surveillance services" by Tommi Meskanen, Valtteri Niemiand Noora Nieminen, provides fascinating insights in the use of advanced cryptography to provide privacy for a surveillance system. The apparent paradox is solved by an innovative way of using garbling, a powerful cryptographic primitive for secure multiparty computation, to achieve privacy-preserving electronic surveillance. The case is assisted living and the client is an elderly person living alone. A security company bases its service on an electronic surveillance system consisting of closed-circuit televisions (CCTV), motion detectors, and/or sensors measuring the activity of the client. The company collects data and analyses the data using data mining, pattern recognition and machine learning tools. The security company has outsourced its data center services into a cloud managed by a third-party company. The data from the surveillance system is stored and analyzed entirely in the cloud environment. Still, our elderly client wants to have his/her privacy and this seemingly impossible goal is what the authors tackle. The garbling techniques themselves are based on secure multiparty computation. The scheme proposed is cutting edge and more research is needed before it becomes practical, but it is refreshing and promising to see these problems being tackled.

The fourth paper, "Cyber security and the Internet of Things: Vulnerabilities, Threats, Intruders and Attacks" by Mohamed Abomhara and Geir M. Køien, is an overview paper. It investigates vulnerabilities in an IoT world, and it looks at various threats and threat types. Both vulnerabilities and threats relate to assets, but what are the assets in an IoT context? Threats do not have to become attacks, but if they do there must be a perpetrator. In security modeling parlance, we often call this entity the intruder. Other questions that need to answered are as follows: What are the attacks? What security goals where there in the first place? Who is the intruder(s)? The old adage Know Thy Enemy basically says that you cannot really win the war unless you understand your enemy.

"It is said that if you know your enemies and know yourself, you will not be imperiled in a hundred battles; if you do not know your enemies but do know yourself, you will win one and lose one; if you do not know your enemies nor yourself, you will be imperiled in every single battle."

Sun Tzu, The Art of War

In a connected world with remote access, you cannot know your enemy by sight, and so the paper discusses these aspects in a generic way, based on observed capabilities and behavior.

Together, these four papers outline different aspects of robustness and resilience for the brave new all-digital world in which IoT/m2m will play significant roles. There is still much research needed, but we hope these contributions will help in the quest to build a safe and secure digital future for us all.

Geir M. Køien,
University of Agder,
Norway

Torrent-Based Dissemination in Infrastructure-Less Wireless Networks*

Kyriakos Manousakis[1], Sharanya Eswaran[1], David Shur[1], Gaurav Naik[2], Pavan Kantharaju[2], William Regli[2] and Brian Adamson[3]

[1]Applied Communication Sciences, Basking Ridge, NJ, USA
[2]Collge of Computing and Informatics, Drexel University, Philadelphia, PA, USA
[3]United States Naval Research Laboratory, Washington, DC, USA
Corresponding Authors: {kmanousakis; seswaran; dshur}@appcomsci.com;
{gn; pk398; regli}@drexel.edu; brian.adamson@nrl.navy.mil

Received 10 February 2015; Accepted 11 March 2015;
Publication 22 May 2015

Abstract

Content dissemination in peer-to-peer mobile ad-hoc networks is subject to disruptions due to erratic link performance and intermittent connectivity. Distributed protocols such as BitTorrent are now ubiquitously used for content dissemination in wired Internet-scale networks, but are not infrastructure-less, which makes them unsuitable for MANETs. Our approach (called SISTO) is a fully distributed and torrent-based solution, with four key features: (i) freedom from any reliance on infrastructure; (ii) network and topology aware selection of information sources; (iii) robust multiple-path routing of content via a proactive peer selection technique; (iv) an integrated distributed content discovery capability, not found in other torrent systems. We have implemented SISTO in software, and evaluated its performance using emulation and realistic mobile network models derived from field measurements. We have measured significant improvements in download latency, resiliency and packet delivery compared to traditional data delivery models and conventional BitTorrent. We have implemented SISTO on both Linux and Android platforms, and integrated it with several android applications for content sharing.

*This work has been funded by the Office of Naval Research (Contract No. N00014-12-C-0377)

Journal of Cyber Security, Vol. 4, 1–22.
doi: 10.13052/jcsm2245-1439.411

Keywords: Mobile ad-hoc networks, peer-to-peer, data dissemination, algorithms, performance, design, reliability, experimentation.

1 Introduction

There exist a wide set of network scenarios, such as first responder and disaster recovery situations, and military and tactical operations, which require applications and protocols to function in a purely ad hoc peer-to-peer fashion, without support infrastructure. (Here support infrastructure is taken to mean the existence of nodes and functionality maintained outside the scope of the set of ad-hoc nodes, yet available to them. The DNS infrastructure is an example of support infrastructure). In many MANET situations, this type of infrastructure support is not feasible. Furthermore, it is critical in such ad-hoc networks, that node or link failures are well tolerated, so that a receiver can obtain data even if the original source is temporarily or permanently disconnected.

While it is possible to address this problem through MANET routing techniques [19], design, development and deployment of routing protocols have long lead times. Consensus in the technical community, on which are the best routing protocols has not emerged despite years of research. Content Distribution Networks (CDNs), such as BitTorrent address the problem above the network layer, thereby avoiding the problem of deploying new routing protocols. In BitTorrent, original content is broken into pieces, and pieces may be individually disseminated. A receiver may obtain pieces concurrently from multiple sources, dispersed across the network. Our approach builds upon the BitTorrent [1] protocol, and addresses the limitations of BitTorrent in a MANET environment as follows:

- Unlike conventional BitTorrent systems, it does not require support from BitTorrent support infrastructure (the servers for content/peer discovery and tracking).
- It includes procedures for identifying and selecting sources based on favorable topological factors and/or network conditions such as congestion in order to form a robust multiple path distribution network for each piece of content. Our multiple path technique is highly tolerant to connectivity disruptions, and offers significantly more robust dissemination than traditional methods, which tend to map all traffic onto a single path between a source and destination.
- It provides for content discovery (not present in BitTorrent and most other CDN technologies, which assume an out-of-band mechanism) using a fully-distributed content discovery mechanism, where in the

users can query for content of interest using key words, and discover the corresponding torrent metadata; by means of the metadata, content is automatically acquired and delivered based on the SISTO torrent algorithm.

The remainder of this paper is organized as follows: Section 2 focuses on related work; Section 3 describes the SISTO architecture and protocols; Section 4 provides a quantitative experimental evaluation of SISTO, and Section 5 provides conclusions.

2 Related Work

Significant related work on peer-to-peer networking exists. Some examples are based on BitTorrent (e.g., SPAWN [5], and CodeTorrent [6], which also uses network coding to help deal with mobile network issues), while others such as 7DS [7], XL-Gnutella [8], are not. In the above related work, the infrastructure question is not addressed, and mechanisms for the discovery of content are not provided – it is assumed that content metadata (e.g., *.torrent* files) are provided through an out-of-band mechanism such as a well-known web server. ORION [9] proposes content query and searching features, but the content technique is does not appear to be separable from the highly specific dissemination technique in that work. In our work, we incorporate a fully distributed mechanism for content discovery based on ProtoSD [13], which is loosely coupled with the rest of the system. In [10], a cross-layer approach for using network information is described, but it does not use peer-to-peer mechanisms, and therefore lacks the required robustness and dissemination efficiency. In [11] a novel torrent-based system is described based on Bluetooth communications, but the mechanisms are coupled with the blue-tooth protocol and do not generalize to other communication techniques. A topology-aware BitTorrent client is developed in [12] for Internet-scale networks in which peers are selected based on hop count and transmission rates. However, it does not take the link quality into account, and uses passive monitoring of connections between peers to estimate the rates. Other work such as [13] and [14] also select peers based on estimating available bandwidth using the technique of packet-pair dispersion. However, as shown in [15], packet dispersion methods do not provide accurate bandwidth estimation in wireless networks. The original SISTO concept was proposed in [16], and in this follow-up paper, we report on the design enhancements of the original concepts and also on results from testing of our fully functioning implementation.

3 SISTO Architecture

When applied to MANETs, the main drawbacks of conventional torrent BitTorrent are (a) that it was developed for large scale, stable networks, and (b) assume infrastructure support, either via peer tracker servers or a core set of servers forming a decentralized peer discovery overlay network. Such assumptions are not suitable for small-scale wireless mobile ad hoc networks (MANETs), especially those that display dynamic behaviour, where connections to any specific servers or infrastructure may not be available, or if available may be subject to frequent disruption.

3.1 Distributed Infrastructure-Free Peer Discovery

In SISTO, like BitTorrent, the peers are discovered using the Distributed Hash Table (DHT) technique in a distributed manner. However the conventional BitTorrent DHT bootstrap process, while distributed, depends on the infrastructure support of the global DHT overlay network. This global DHT network assumed by BitTorrent is "infrastructure" for torrent systems in the same way as the DNS server overlay is infrastructure for IP networks. Furthermore, because of the scale of the global DHT overlay network support infrastructure, in conventional BitTorrent, a new peer needs to connect to at least 50 other peers in order to participate in the DHT network. In a typical MANET, there may not even be 50 nodes in the entire network. In SISTO we have modified the bootstrapping by: a) removing the dependencies on the global DHT network, and b) we have set the number of DHT peers to be a configurable parameter depending on the characteristics of the corresponding network (e.g., size). In SISTO, a peer bootstraps from a local ad-hoc DHT overlay network, using a set of known locally stored addresses that have been configured manually or learned from previous peer connections. This enhancement allows SISTO use the members of the torrent swarm to form a peer discovery DHT network among themselves, allowing it to operate without any connectivity to the torrent DHT support infrastructure in the Internet. This seemingly minor alternation in the architecture completely changes the applicability of this approach, making it possible to be used in MANETs. **Figure 1** depicts the enhanced, fully distributed infrastructure-less DHT bootstrap process in SISTO.

Since the DHT discovery network in SISTO will now be typically a much smaller (MANET) sized entity, we take advantage of this fact to speed-up peer discovery, by introducing a new peer discovery control parameter, which can be set to allow an earlier start of downloads.

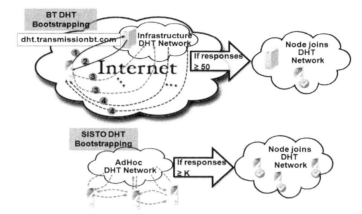

Figure 1 DHT bootstrapping in SISTO vs. BitTorrent.

3.2 Network-Aware Peer Selection

Peer selection is the process by which each peer decides which subset of its peers to upload data to. In SISTO, we have made another change relative to conventional BitTorrent where peers are selected in a random fashion. In SISTO, peers are selected using cross-layer information in a network- and topology-aware manner. Furthermore, the number of peers (called the Upload Number in this paper, which is normally fixed in BitTorrent) is adjusted dynamically based on network conditions. The network-aware peer selection is designed with the following objectives: (i) reducing long distance transmissions and localizing the transmissions, so that the channel contention and interference is reduced, (ii) utilizing stable, high performing links so that the efficiency of data dissemination is higher, and (iii) since it is common for such networks to have links with frequently varying bandwidth resulting from factors such as mobility and terrain effects, avoiding the problem of underutilization of low bandwidth or mildly lossy links (which are treated unfairly with the conventional BitTorrent scheme).

The cross-layer information needed by SISTO can be acquired in multiple ways. Firstly, SISTO includes a simple Network Monitoring tool, enabling each node to periodically gather latency, hop count and loss information with respective to other known peers in the swarm. SISTO is designed to read this information from a pre-specified *location:port* and in a self-descriptive, standardized JSON format. This architectural decision allows any third-party process that provides network information, such as a routing protocol agent, Dynamic Link Exchange Protocol (DLEP) agent or a dedicated network awareness service [23], to be easily plugged into the SISTO interface in place of the SISTO-provided tool.

SISTO implements the following three algorithms for network-aware peer selection (PS):

Hop-only PS: The peers are ordered based on the hop count from the uploading node, preferring the closer peers. When there is a tie, the download rates (which are used in conventional BitTorrent – referred to as TitForTat or TFT) are used as secondary criteria.

Latency-Hop PS: The peers are ranked in the increasing order of latency (round trip time) between the peer and the uploading node. When there is a tie, the hop count and download rates are used as secondary and tertiary criteria.

Loss-Hop PS: The peers are ranked in the increasing order of packet loss between the peer and the uploading node. When there is a tie, the hop count and download rates are used as secondary and tertiary criteria.

Once the peers are ranked using one of these policies, the data is uploaded to the top N peers, where N is the upload number, and the other peers are choked. This peer selection process (called the re-choke cycle) is repeated every 10 seconds. It is assumed that the Network Monitoring tool described above is providing hop count, latency and loss measurements.

Another important factor that impacts performance is the value of the upload number itself, since if it is too low, the throughput achieved is low, and if it is too high, network congestion may result. Unlike BitTorrent, SISTO adjusts the upload number *dynamically*, based on the network conditions. The re-evaluation of upload number occurs every alternate (configurable) re-choke cycle, i.e., every 20 seconds (by default) in our implementation. The parameters that are used to determine the upload number value are (i) total upload rate (TUR) across all peers, i.e., the total number of bytes per second that the node uploads to all its peers averaged over the sampling time window and (ii) average latency (AL) between the node and its peers. When it is time to evaluate the upload number at a node, the current TUR and AL are compared with the TUR and AL from the previous cycle. If there is an increase in the AL or a drop in TUR, it indicates the build up of congestion in the network. If there is a decrease in AL or increase in TUR, it indicates the availability of network capacity, especially when new peers join the network. Based on these observations, a heuristic adjustment policy is employed, as shown in **Table 1** the upload number is linearly incremented when AL decreases or TUR increases, linearly decremented when AL increases or TUR decreases, and unchanged in other cases.

Table 1 Dynamic adjustment policy

AL	TUR	Upload Number
↑	↓	−1
↓	↑	+1
↑	↑	+0
↓	↓	+0

3.3 Adaptive Re-Routing via Proactive Peer Creation (PPC)

In conventional BitTorrent, peers that serve as seeds are selectively reactively, based solely in whether or not an application attached to that peer wants to receive the information in question. Proactively selecting new seeds enables a highly robust dynamic route selection at the application layer. For example, suppose there is a single source-receiver pair (*n5* and *n8*) in the network, as shown in **Figure 2**. Suppose that the shortest path between the two nodes, as chosen by conventional routing, is of poor quality, possibly due to congestion. Suppose that there exists an alternative, but longer path between *n5* and *n8*, which is not congested. If we trigger one of the nodes in the alternative path, say *n12*, to become an additional seed for the torrent, then n5 sends data to *n12*, and *n8* subsequently receives it from *n12*. The torrent algorithm will then naturally begin to favor the high-performing path. In other words, the torrent dissemination process steers the traffic through the alternate, better route. SISTO exploits this potential and defines mechanisms to identify such peers, which can improve performance significantly. Furthermore, enabling more than one such peer can cause multiple new paths between the source

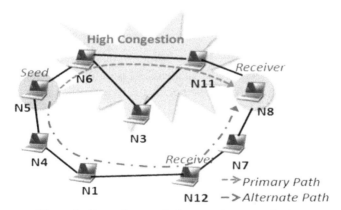

Figure 2 Example scenario for proactive peer selection.

and the destination to emerge. Thus traffic can be disseminated to receivers on multiple paths, which offers not only performance improvements, but makes the entire dissemination process more robust, since multiple paths may be active in parallel, and information dissemination adaptively favors the better performing paths. Note that that can also occur in conventional BitTorrent, but without PPC, if it occurs it is dependent on other peers in the right place being interested in receiving the content, while PPC ensures that it happens by design. We believe that there is a high potential in proactive peer creation, especially in congested networks, when transient links become available, and in the vicinity of "weak spots" in the network, (which are nodes that if they fail, will cause partitioning).

In the current version of SISTO, we design and implement PPC to address the aforementioned problem of congestion. Accordingly, when the observed latency between a peer and its actively downloading peer (i.e., peer that it uploads data to) exceeds a threshold value, PPC is enabled for this source and receiver pair (S, R) and torrent T. The source peer (i.e., the uploading peer where PPC is enabled) obtains a list of known peers in the network by reading the DHT overlay node list (we assume that all peers want to participate in PPC). These peers may or may not already be a member of torrent T's swarm (since the DHT network is established independent of the torrents being exchanged). Let this list of peers be $L_1 = \{p_1, p_2, p_3, ...\}$. S then obtains the latencies between S and each node in L_1 from the local Network Monitor. Let this list of latencies be $L_2 = \{lat(S, p_1), lat(S, p_2), lat(S, p_3), ...\}$, where $lat(a, b)$ is the latency between peers a and b. The nodes in L_1 are sorted in the increasing order of latencies from L_2. A fraction of nodes in the sorted list (50% in our implementation), are queried for their observed latencies to R. Let this list be $L_3 = \{lat(p_1, R), lat(p_2, R), lat(p_3, R), ...\}$. The number of nodes queried can be changed according to the desired trade off in communication overhead. Based on L_2 and L_3, the node p_i which yields the least total latency $lat(S, p_i) + lat(p_i, R)$ is selected to be enabled as a peer for T. A request is sent to p_i to add the torrent; if the new node rejects the request, then the node with the next lowest total latency is selected. This algorithm is highly adaptive to changing network conditions.

3.4 Content Discovery

In most conventional CDNS (including BitTorrent) content discovery is assumed to take place out of band (e.g., via email, or social media). In SISTO, an application can request content either using a set of keywords, by selecting content from keyword search results, or by directly referencing specific content metadata. The Content Discovery component is responsible for creating metadata and distributing these across the network, as well as publishing advertisements from peers that want to seed and share content. Nodes that are interested in the published content use the relevant metadata obtained in the content discovery process, and then discover peers by means of DHT techniques. Once a torrent swarm of peers is established for the requested content, data dissemination begins.

For content metadata, SISTO uses magnet links instead of the traditional ".*torrent*" metafiles, because of their small size. The small size of the metadata associated with the torrent content, allows it to be disseminated using very simple techniques, i.e., the magnet links are pushed onto the network and discovered by peers in a distributed manner, using ProtoSD [13], which is a discovery system, which helps publish, query and disseminate content references across the network. ProtoSD uses two service discovery protocols: (1) multicast-DNS (MDNS) [4] and (2) Independent Discovery Interface (INDI) [12]. Using INDI, ProtoSD is able to discover and disseminate services effectively on dynamic, low-connectivity networks without any infrastructure, similar to the modified DHT technique described above.

SISTO allows content-providing peers (seeds) to tag content with keywords and other metadata, and push content advertisements to the network periodically, which are picked up by other nodes and stored in their local knowledge base, as shown in **Figure 3**. A sample advertisement of media content is shown in **Figure 4**. Along with advertising, the nodes also seed their content, i.e., makes the content available to other peers as a torrent for

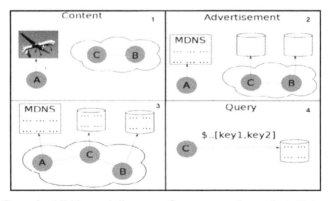

Figure 3 Content publishing and discovery: Content created at node A (1) is advertised to other nodes (2, 3). If a query fails in the local knowledge base, it is retrieved from a different node (4).

SquadAEnemySniperSpotted,_http._tcp.local
.,local,
192.168.141.132/8080 originating from:
192.168.141.132, text = ProtoServiceID=0,
gps-coordinates=["10,10"!],
DiscoverySystem=INDI, keywords={!
"squadA1": "", !
"position": "", !
"EnemyBase": "", !
"snipers": "", !
"enemies": ""!
}, _service_addr=192.168.141.132,
metadata={!
"topic": {"squada": ""}, !
"timestamp": {"1389284174": ""}, !
"author": {"emu0": ""}, !
"mime": {"image/png": ""}, !
"filename": {"squad_a.jpg": ""}!
}, _producer_int=192.168.141.132,
magnetLink={!
"magnet":{"magnet:?xt=urn:btih:bc9ae647a3
e6c3636de58535dd3f6360ce9f4621"}}

Figure 4   Sample content advertisement.

download. A client that wants to download content can query for keywords pertaining to the relevant magnet link. If the node that pushed the link advertisement fails, the magnet link may be retrieved from the knowledge base of other nodes. The node subsequently uses this magnet link to download the data via the torrent algorithm.

4 Evaluation

The SISTO system has been implemented in C++ building from the libtorrent library [20]. The software implementation has been evaluated using a realistic 30 node mobile network obtained from field measurements emulated on a Common Open Research Emulator (CORE)/Extendable Mobile Ad-hoc Network Emulator (EMANE) testbed [21, 22]. Both static and mobile versions of this network are used in the experimentation. For the experiments in Section 4.1 below, we used the basic range model (130 m) with the bandwidth on all the links set to 200 Kbps, an average packet loss of 5% and delay of 20ms. For the experiments in Sections 4.2, 4.3 and 4.4, the CORE testbed, which emulates the network layer and upper layers, was integrated with

Extendable Mobile Ad hoc Network Emulator (EMANE) for emulating lower layers (e.g., the 802.11abg MAC model was applied), and the link rates were set to 2 Mbps.

4.1 Peer Discovery

We compare the performance of SISTO's enhanced DHT with BitTorrent's DHT by downloading 5 video files across the network (in both static and mobile scenarios); each file size was 4.5 MB, and was seeded by 1 peer and requested by 10 peers. The experiments were repeated with different sets of seeds and receivers that were selected randomly. **Figure 5** shows the average peer discovery latency in SISTO and BitTorrent. We observed that SISTO reduces the peer discovery latency by 7.7% on average. To see the impact of these differences on the actual data download, we measured the download latency, i.e., the average time taken to for a receiver to download the entire file. **Figure 6** shows the download latency for BitTorrent and

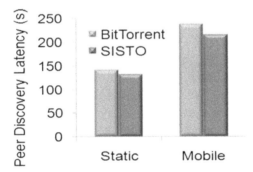

Figure 5 Average peer discovery latency of SISTO vs. BitTorrent.

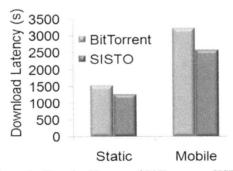

Figure 6 Download latency of BitTorrent vs. SISTO.

SISTO (without network/topology awareness). We see that SISTO reduces the download latency by 19% on average. The overhead of peer discovery in SISTO, i.e., the overhead of SISTO's enhanced DHT messaging, as a fraction of the total network traffic is only a small fraction of the total network traffic (<1.5%) and scales well with network churn.

4.2 Content Dissemination Efficiency

Similar sets of experiments described in Section 4.1 were used to evaluate the content dissemination efficiency of SISTO. **Figure 7** shows the download latency for experiments in both static and mobile network scenarios. The performance of SISTO is compared with conventional HTTP, which relies on the underlying routing protocol (OLSR in this case) for failure recovery.

It may be noted that the network-aware optimizations were not enabled in SISTO for these experiments. We see in **Figure 7** that even in a static network only 55% of the file downloads actually completed – this is due to the conventional approach's limited ability to recover from mobility driven degradation. Furthermore, in the mobile network, even less, namely 32% of downloads completes. In contrast, SISTO delivered 100% of the requested content both static and mobile cases, indicating that it is much more robust to both congestion and mobility effects.

Figure 8 shows the measured the traffic overhead incurred by SISTO control messages for data dissemination, such as peer handshake, piece request, etc. We observe that this is a small fraction of the total traffic (<1%), and also that the overhead is lower in 5-torrent than 1-torrent, indicating scalability of SISTO.

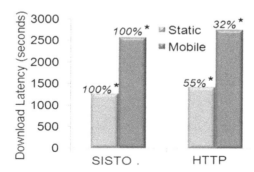

Figure 7   Download completion (HTTP vs. SISTO).

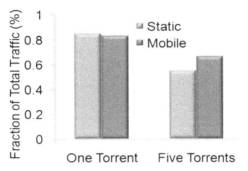

Figure 8 Content dissemination overhead of SISTO.

4.3 Network-Aware Peer Selection

In this series of experiments, we employed the same five-torrent scenario described above and evaluated the SISTO network-aware peer selection algorithms against each other, and against tit-for-tat (TFT) peer selection of BitTorrent. A static network without any node mobility was used, and the upload number of peers was fixed to 3.

Figure 9 shows the average download latencies. We see that the network-aware peer selection yields better performance than TFT. For this scenario, the use of the Hop-only scheme reduced the download latency by 19% and Latency-Hop by 28% compared to TFT. Note that although Latency-Hop performed the best in this experiment, there may be other scenarios or situations where the other network-aware peer selection algorithms are useful. E.g, when some links are lower bandwidth than others, Hop-only prevents starvation of these links; or when only loss or hop measurements are available, Loss-Hop or Hop-only algorithms may be selected.

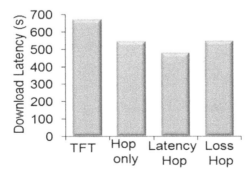

Figure 9 Download latency with different peer selection algorithms.

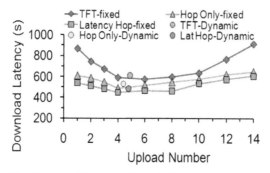

Figure 10   Download latency with and without dynamic adjustment.

Next, the dynamic adjustment of upload number was evaluated. **Figure 10** shows the average download latencies for TFT, Hop-only and Latency-Hop, for various fixed values of upload number (N). Note that N significantly impacts performance. Next, the dynamic adjustment algorithm is enabled, and the experiments are repeated. The average value of N over the course of the experiment was computed. We see that the dynamic adjust algorithm tends to converge to the optimal region of the upload number.

4.4 Proactive Peer Creation (PPC)

PPC was evaluated in CORE/EMANE using 5 seeds, each with one receiver, and with a file of size 4.5 MB. The seeds were chosen randomly and the receivers are chosen such that they were at least two hops away from the corresponding seeds. Cross-traffic at the rate of 750 Kbps was introduced in the routes between each seed-receiver pair, using IPerf UDP [3]. Five runs of the experiment were conducted, each with a different set of seeds and receivers. The latency threshold was set to 4000ms (exceeding that level triggers PPC).

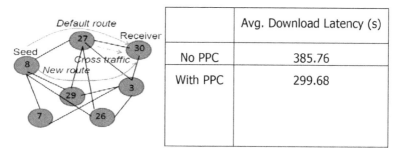

	Avg. Download Latency (s)
No PPC	385.76
With PPC	299.68

Figure 11   PPC experimental setup and latency.

In the scenario of **Figure 11**, when PPC was activated, node *n29* was found to yield the least latency path between *n8* and *n30*, and hence was enabled as a peer. Subsequently, *n30* received most of its data via *n29*. Similarly, the other seed-receiver pairs enabled new peers as needed. **Figure 11** shows the average download latency per receiver with and without PPC. We see that PPC reduces the download latency by about 22%. Furthermore, the percentage of retransmissions in the network was 15.2% without PPC, but was reduced to 9.6% with PPC. The additional overhead due to the enabling of new peers was measured in terms of the number of control bytes per data byte (i.e., ratio of torrent control bytes to file size), which was 0.003 without PPC, and 0.007 with PPC.

5 Conclusions

We have developed a distributed torrent-based file distribution suite that operates efficiently without any infrastructure support. It is agnostic to underlying network routing protocols, and includes a fully decentralized content discovery component. Our evaluation shows that SISTO is significantly more robust both than conventional routing and infrastructure-based CDN approaches, and the addition of two key elements, i.e., adding network-awareness to peer selection, and proactive instantiation of peers to enrich routing diversity further improves robustness and throughput. We have also prototyped SISTO on an Android platform, and tested it in smart-phone networks envisaged for disaster-recovery, tactical or first responder networks.

References

[1] B. Cohen, "Incentives Build Robustness in BitTorrent", P2PECON03.
[2] P. Maymounkov et al: "A Peer-to-peer Information System Based on the XOR Metric", IPTPS, 2002.
[3] https://code.google.com/p/iperf/
[4] S. Cheshire et al., *Multicast DNS*. IETF RFC 6762.
[5] S. Das, et al, "SPAWN: A Swarming Protocol For Vehicular Ad-Hoc Wireless Networks", VANET 2005.
[6] U. Lee et al, "CodeTorrent: Content Distribution using Network Coding in VANET", MobiShare 2006.
[7] S. Srinivasan et al, "7DS - Node Cooperation and Information Exchange in Mostly Disconnected Networks, IEEE ICC 2007.

 [8] M. Conti et al, "A cross-layer optimization of gnutella for mobile ad hoc networks", MobiHoc 2005.
 [9] A. Klemm et al, "A Special-Purpose Peer-to-Peer File Sharing System for Mobile Ad Hoc Networks, IEEE VTC 2003.
[10] M. Schurgot, et al, "Providing local content discovery and sharing in mobile tactical networks", IEEE MILCOM, 2013.
[11] Sewook Jung, et al. "BlueTorrent: Cooperative Content Sharing for BluetoothUsers", IEEE Percom, White Plains, NY, USA, March 2007
[12] S. Ren, E. Tan, et al, "TopBT: A topology-aware, infrastructure-independent BitTorrent client", IEEE INFOCOM 2010.
[13] M.K. Sbai, et al, "P2P content sharing in spontaneous multi-hop wireless networks", COMSNETS, 2010.
[14] M. Kawarasaki, "Network-aware peer selection method for P2P file downloading using packet-pair measurement", IEEE ICUMT, 2011.
[15] J. Sucec, et al. "A resource friendly approach for estimating available bandwidth in secure mobile wireless networks", MILCOM, 2005.
[16] D. Shur et al, "SISTO: A proposal for serverless information services for tactical operations", IEEE MILCOM 2012.
[17] J. Macker, I. Taylor, "INDI: Adapting the multicast DNS service discovery infrastructure in mobile wireless networks", IEEE MILCOM 2011.
[18] http://www.nrl.navy.mil/itd/ncs/products/protosd
[19] A. Shrestha and F. Tekiner, "On MANET Routing Protocols for Mobility and Scalability", International Conf on Parallel and Distributed Computing, 2009.
[20] *Libtorrent,* www.libtorrent.org
[21] Ivanic, N et al. "Mobile ad hoc network emulation environment." *MILCOM 2009.* IEEE, 2009.
[22] Ahrenholz, J. et al. "Integration of the core and emane network emulators." *MILCOM 2011.* IEEE, 2011.
[23] Chen, Ta, et al. "Enhancing application performance with network awareness." *MILCOM 2011.* IEEE, 2011.

Biographies

K. Manousakis is a senior scientist with Applied Communication Sciences (ACS) since 2005. He has earned his MSc and PhD degrees from the Electrical and Computer Engineering Department of University of Maryland, College Park in 2002 and 2005, respectively. His expertise is in the areas of network protocols and network optimization. His recent interests involve content distribution for mobile networks, cognitive radios networking and interference mitigation. He has been a primary investigator (PI) for many US government funded programs like the ARL Collaborative Technology Alliance (Advanced Structures for Tactical Networks), the CERDEC NetMining (Dynamic Network Management for MANET), CERDEC Network Design (Offline and real time network planning tool) and the DARPA Fixed Wireless At a Distance (Content Distribution in Challenging Environments). Dr. Manousakis has patents and numerous publications in high profile conferences, journals and magazines in the areas of optimization techniques and algorithms for resource constrained mobile wireless networks. He has organized and chaired multiple IEEE sponsored conferences. Dr. Manousakis is the recipient of Award of Excellence in Telecommunications by Ericsson and in 2009 he has been awarded the IEEE PCJS Leadership Award.

S. Eswaran is a Research Scientist at Xerox Research Center India. Previously, she was a Senior Research Scientist at Applied Communication Sciences at Basking Ridge, NJ, where she worked on several research topics in the field of wireless and mobile networking. She received her Ph.D. degree in Computer Science and Engineering from Penn State in 2010, and M.S in Computer Engineering from Univeristy of Virginia in 2006. Her research focuses on optimal and robust data delivery across resource-constrained and variable wireless networks.

D. H. Shur received his Ph.D and MS degrees in Electrical Engineering from Stanford University in 1987 and 1981, respectively. From 1986–1996, he was with AT&T Bell Labs, Holmdel, NJ, where he was a Distinguished Member of Technical Staff. From 1996–2003, he was with AT&T Labs-Research, Middletown, NJ, where he was a Technology Consultant. From 2003–2008, he was with Bloomberg LP, New York, NY, where he was an Infrastructure Systems Architect. Since 2008 he has been with Telcordia Technologies/Applied Communication Sciences, where he is currently a Chief Scientist in the Mobile Networking Department. Dr. Shur has been the Principal Investigator for a number of research programs funded by DARPA, the US Office of Naval Research, and the US Army Communications-Electronics Research, Development and Engineering Center (CERDEC). His research has spanned the areas of IP-TV convergence (and was standardized in DOCSIS 3.0), real-time multimedia, IP over ATM, Packet Switching technologies, Electronic Watermarking, Web Caching, Multimedia Communication, Wireless Systems, Distributed Systems, Data Distribution, and Circuit Network/Packet Network convergence. He received the best paper award at IEEE WOWMOM in 2010.

G. Naik is a Senior Research Scientist with the Applied Informatics Group (AIG) at the Drexel University College of Computing & Informatics. His research interests in applied R&D include software defined networks, cyber security, and mobile computing. He has contributed to a number of open source projects and to community standardization working groups. Gaurav holds a M.S. degree in Computer Engineering.

P. Kantharaju is a graduate student studying Computer Science at Drexel's College of Computing and Informatics (CCI). He has a B.S. in Computer Science from Drexel University and his research interests include distributed computing and peer-to-peer networking.

W. Regli is a Professor (on leave) of Computer and Information Science in the Drexel University College of Computing & Informatics. He holds departmental appointments in Mechanical Engineering and Mechanics (College of Engineering); Electrical and Computer Engineering (College of Engineering); and in the College of Biomedical Engineering, Science and Health Systems. From 2007–2010 Regli was Senior Science Adviser to the National Institute of Justice's Communications Technologies Center of Excellence. He currently (2010–) serves as Senior Scientific Adviser to the Defense Programs Office of the National Nuclear Security Administration of the U.S. Department of Energy in areas of information technology for design, manufacturing, production in support of the stewardship and surety of the United States' nuclear weapons stockpile.

B. Adamson has been involved in research in radio communications and data networking at the Naval Research Laboratory (NRL) since 1984. His background includes digital signal processing and spread spectrum communications. Mr. Adamson's research interests include data and multimedia network transport, group communications, dynamic routing for wireless networks, and peer-to-peer networking. He has been focused on data networking for wireless and other networks for the past twelve years. This has included active participation in the Internet Engineering Task Force (IETF) in the areas of IPv6, reliable multicast transport, and Mobile Ad-hoc Networking (MANET). He also serves as co-chair of the Internet Research Task Force (IRTF) Network Coding Research Group (NWCRG). Mr. Adamson is also the principal author of several different network protocol implementations and network test, analysis, and visualization tools that have been applied in DoD, government, academic and commercial research and development, demonstration, and operational use.

Cyber Security for Intelligent World with Internet of Things and Machine to Machine Communication

Vandana Rohokale and Ramjee Prasad

Center for TeleInFrastruktur (CTIF), Aalborg University,
Aalborg, Denmark
Corresponding Authors: vmr.301075@gmail.com; Prasad@es.aau.dk

Received 5 February 2015; Accepted 25 February 2015;
Publication 22 May 2015

Abstract

The growth of interconnected objects through Internet of Things (IoT) and Machine to Machine Communication (M2M) is no doubt inevitable. The researchers have predicted that by 2020, around fifty billion objects throughout the world will be connected with each other with the help of internetwork of smart objects. With the number of networked objects, grows the number of cyber-crime threats. Forthcoming fifth generation of mobile communication will be the converged version of all the wired and wireless networking services and technologies. The heterogeneous networking approach gives rise to various cyber threats. Design of robust cyber security solutions for such heterogeneous networks with smart devices is challenging task.

Keywords: Internet of Things (IoT), Machine to Machine Communication (M2M), Cyber Security, Cybercrime, etc.

1 Evolution of Internet

Internet first generation dates back to 1970's when the Advanced Research Projects Agency Network (ARPANET) introduced first Internet service which was intended for Military, Government and Educational Institutes in United

Journal of Cyber Security, Vol. 4, 23–40.
doi: 10.13052/jcsm2245-1439.412

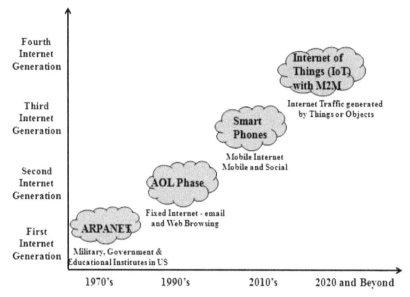

Figure 1   Internet evolution from ARPANET to IoT and M2M.

States. Following that, America Online (AOL) second phase emerged which gave birth to fixed internet that facilitated email and web browsing during 1990's. Current third phase that is 2010's is the era of smart phones with mobile internet experience which is faster and better. Now the whole world is looking forward for the revolutionary fifth generation (5G) of mobile communication. Internet of Things (IoT) and Machine to Machine Communication (M2M) are the integral part of 5G. So the fourth generation of Internet is termed as Internet of Things or objects where most of the traffic will be generated by the interaction of smart objects of the physical world with the digital world [1]. Figure 1 depicts the evolution of Internet from ARPANET to IoT and M2M. On global scenario, per person at least five gadgets are considered to be networked and as a whole, the count of networked objects may go beyond 50 Billion.

2 Internet of Things (IoT) and Machine to Machine Communication (M2M)

M2M comes under the huge umbrella of IoT. There are different opinions among researchers about IoT and M2M. It is becoming the debatable question like egg first or hen first. But these two concepts are technically very much different. M2M is the communication protocol for the interactions among machines with intelligence, or machine to human interface. IoT is the networking service which enables interworking of all such smart machines. For example, credit or debit card of the bank ATM is the example of M2M because the ATM machine reads the information on the card and acts according to the requirements of user. But when user leaves the bank, automatically the fans and lights are off, which is the example of IoT where in, the human presence is detected and accordingly the electrical appliances are turned on or off. In essence, M2M is the plumbing for strong connectivity among network objects for best networking experience in IoT [2].

The Internet of Things is expansion of the current Internet services for each and every object which exists in this world or likely to exist in the coming future. As IoT has become an active research area, different methods from various points of view have been explored to promote the development and popularity of IoT [3]. One trend is viewing IoT as Web of Things where the open Web standards are supported for information sharing and device interoperation. While bringing smart things into existing web, the conventional web services are needed to be enriched and integrated with the physical world [4]. The IoT is rapidly gaining much of the attention in the scenario of modern wireless telecommunications. The basic idea of the concepts such as Radio Frequency Identification (RFID) tags, sensors, actuators, mobile phones, etc., which, through unique addressing schemes, are able to interact with each other and cooperate with their neighbours to reach common goals [5].

Actually, many challenging issues addressed on both technological as well as social knots have to be untied before the IoT idea is widely accepted. Central issues must try to support possible interoperability of interconnected devices, providing with higher degree of smartness by enabling their adaptation and autonomous behaviour, while guaranteeing trust, privacy, and security. Also, the IoT idea poses several new problems concerning the networking aspects. In fact, the things composing the IoT will be characterized by lesser requirement of resources in terms of computation and energy capacity. Accordingly, the proposed solutions need to pay special attention to resource efficiency besides the obvious scalability problems. The Internet of Things, an emerging global Internet-based technical architecture facilitating the exchange of goods and services in global networks has an impact on the security and privacy of the involved stakeholders. Measures ensuring the architectures for data authentication, access control and client privacy need to be established.

M2M means Machine to Machine communication which is the most popular interface for Internet of Things (IoT) in the present mobile wireless communication, whose data transmission is supported by cable, wireless, mobile and other technologies, which may suffer from significant security vulnerabilities and risks. M2M is widely used in power, transportation, industrial control, retail, public services management, health, water, security and other industries. M2M is typically required to be small, inexpensive and those able to operate unattended by humans for extended periods of time and to communicate over the wireless area network. It can achieve vehicle theft security, safety monitoring, auto sales, mechanical maintenance, and public transport management, logistic and other functions. The most important part in IoT Internets is the connection and interoperability between machines, which is called M2M. Security services such as data integration, authentication and key establishment are critical in M2M [6, 7]. Figure 2 shows various machine intelligence perspectives necessary for the IoT with the help of M2M.

There are great prospects of development and applications in IoT, which can be applied in almost every aspect of human life, such as environmental

Figure 2   Machine intelligence perspectives for IoT through M2M with Cyber Security.

monitoring, medical treatment and public health, Intelligent Transportation System (ITS), smart grid and other areas. Main enabling factor for promising paradigm in the integration of several technologies and communications solutions is the IoT. Identification, sensing and monitoring technologies, wired and wireless sensor and actuator networks, enhanced communication protocols and distributed intelligence for smart objects are just the most relevant. Table 1 shows some of the recent news related to IoT and M2M Communication.

Table 1 Recent news related to IoT and M2M communications

News Headline	Description	Ref
The upcoming boom in smart buildings	After several years of slower than expected growth, the smart building sector is poised to skyrocket. IDC expects the market to triple to $21.9 billion in just four years.	[8]
Smart grid advice from SMUD (Home gateways? Really?)	SMUD's Smart Sacramento smart grid program deployed 615,000 smart meters. The AMI network also connected to home area network (HAN) devices. SMUD deployed 6,700 HAN devices. The project was funded in part by a $127.5 million grant from the Department of Energy. News Date: 2014-7-21	[9]
Smart-Grid Sensor Market Steadily Climbing to $39 Billion by 2019	The market for all types of sensors used in smart grid applications will grow from $26.4 billion this year to $36.5 billion in 2019 and nearly $46.8 billion by 2021, according to a new report from industry analyst firm NanoMarkets. News Date: 2014-9-15	[10]
Electric scooter with Smartphone connection is launched in Europe by Terra Motors	Terra Motors Corporation, Japan's manufacturer of electric two-wheelers and three-wheelers, starts commercial sale of "A4000i" in European countries, the electric scooter with smartphone connection. News Date: 2014-9-12	[11]
Smart connected homes driving IoT	The Internet of Things is set to create a market worth almost US$9 trillion by 2020. According to some analysts, a large chunk of this business will be generated by smart connected homes. Antony Savvas reports on business developments in the smart home sector. News Date: 2014-9-1	[12]
Advancing LTE migration heralds massive change in global M2M modules markets	A central element of research for this report involved extensive interviews with 20 modules manufacturers, mobile network operators,	[13]

(Continued)

Table 1   Continued

News Headline	Description	Ref
	M2M service providers and M2M device manufacturers in order to assess current pricing trends for the 13 most common variants of M2M modules available in the global market. News Date: 2013-12-22	
Technology migration strategies of US carriers will trigger a new era of LTE for the M2M industry	The combination of falling module pricing and the high costs of replacing legacy modules will fundamentally change the technology planning and cost analysis of long-term M2M deployments. News Date: 2013-12-18	[14]
The connected car: a US$282 billion opportunity, but who pays?	The automotive sector is probably the most exciting in M2M, particularly for mobile network operators, module vendors, and sundry others associated with cellular M2M. Machina Research forecasts that by 2022 there will be 1.5 billion automotive M2M connections globally, up from 109 million at the end of 2012. News Date: 2013-12-18	[15]
The rise of M2M/IoT Platforms highlights new commercial dynamics, and new challenges	Any potential provider of M2M/IoT Application Platform solutions must move significantly beyond this core capability in order to attract application developers and build the ecosystems necessary to gain critical mass. News Date: 2013-12-17	[16]
Berg Insight says 2.8 million patients are remotely monitored today	According to a new research report from the analyst firm Berg Insight, around 2.8 million patients worldwide were using a home monitoring service based on equipment with integrated connectivity at the end of 2012. News Date: 2013-12-15	[17]
M2M: The focus is still on people	The idea of machines speaking to machines makes some people worry that there is an impending machine age. In reality, machine-to-machine (M2M) communication is not science fiction, but science fact and, says Daryl Miller of Lantronix, human beings are being anything but left out. News Date: 2013-12-11	[18]
V2V penetration in new vehicles to reach 62% by 2027, according to ABI Research	Vehicle-to-vehicle technology based on DSRC (Dedicated Short Range Communication) using the IEEE 802.11p automotive Wi-Fi standard will gradually be introduced in new vehicles driven by mandates and/or automotive industry initiatives, resulting in a penetration rate of 61.8% by 2027. News Date: 2013-12-10	[19]

3 Security and Privacy Issues in IoT

With the existence of IoT, users will be surrounded and tracked by thousands of smart objects. Security and privacy of the user is of utmost importance. There are various issues related to these parameters such as data confidentiality and trust negotiations which are discussed below.

Dynamicity and Heterogeneity: IoT is the most diverse network where many devices will be added and removed from the network on the go. Privacy and security provision for such diverse and heterogeneous network is the great challenge.

Security for Integrated Operational World with Digital World: Control planes designed till date have not considered security provisions. But the integration of physical and digital world with internetwork connectivity demands security.

Data Security with Device Security: Lot of research work has been contributed in the direction of device security. Now is the time for data security with device security. IoT and M2M aims at communication among objects which demands data security.

Data Source Information: It is critically important to know that from where the data has originated. Knowledge about data source is very important for control, audit, manage and ultimately secure the IoT and M2M communication [27].

Data Confidentiality: Data confidentiality represents a fundamental issue in IoT scenarios, indicating the guarantee that only authorized entities can access and modify data. This is particularly relevant in the business context, whereby data may represent an asset to be protected to safeguard competitiveness and market values. In the IoT context not only users, but also authorized objects may access data. This requires addressing two important aspects: first, the definition of an access control mechanism and second, the definition of an object authentication process (with a related identity management system).

Trust negotiation: The concept of trust is used in a large number of different contexts and with diverse meanings. Trust is a complex notion about which no consensus exists in the computer and information science literature, although its importance has been widely recognized. Different definitions are possible depending on the adopted perspective. A main problem with many approaches towards trust definition is that they do not lend themselves to the establishment of metrics and evaluation methodologies.

A widely used security policies are for regulating accesses to resources and credentials that are required to satisfy such policies. Trust negotiation refers to the process of credential exchanges that allows a party requiring a service or a resource from another party to provide the necessary credentials in order to obtain the service or the resource. This definition of trust is very natural for secure knowledge management as systems may have to exchange credentials before sharing knowledge. For this reason, we base our analysis of trust issues in IoT upon it. Trust negotiation relies on peer-to-peer interactions, and consists of the iterative disclosure of digital credentials, representing statements certified by given entities, for verifying properties of their holders in order to establish mutual trust. In such an approach, access resources are possible only after a successful trust negotiation has been completed [28].

4 Security in Machine to Machine Communication

M2M devices will ultimately connect to core network services through a variety of means, from direct broadband or capillary wireless networks, to wired networks [29]. Capillary networks used by M2M systems are made of a variety of links, either wireless or wired. Network's role is to provide a more comprehensive interconnection capacity, effectiveness and economy of connection, as well as reliable quality of service. Because of the large number of nodes in M2M, it will result in denial of service attacks when data spread, since a large number of machines sending data results into network congestion. In the service network, an attacker may eavesdrop user data, signalling data and control data and unauthorized access to stored data within the system network elements, or do passive or active flowing analysis.

An attacker through the physical layer or protocol layer can interfere in the transmission of user data, signalling data or control data may use network services to impersonate legitimate users, or take advantage of posing access to legitimate users by pretending services network to access network services, to obtain unauthorized network services. To prevent unauthorized access to services in Remote Validation processes, the relying party directly assesses the validity of the device based on the evidence for the verification received. Local verification is only passive, just measuring integrity values of the loaded and started components.

Attackers often access, modify, insert, delete, or replay the user communication information by physical theft, online listening, posing as legitimate users and other means, such as the Man-in-the-Middle (MITM) attacks, which can steal or change the course of M2M communication of information between devices in the process of "intercept data-modification of data – sending data", resulting in the loss of legitimate users [30]. Typically, to obtain certain confidential information, an attacker will obtain communications data in any ways, such as the use of online listening, MITM attacks and other. Hence the data communication between M2M devices needs to be integrity and confidentiality protection, and M2M equipment should have appropriate mechanisms to accomplish this function.

An attacker may access the applications software and signing information for M2M by malware, Trojans or other means, and then replicate in other M2M devices to restore to fraudulent use of M2M identity of the user; he also can change, insert, and remove the user's communications data by a virus or malicious software. Anti-virus software applications will reduce viruses and malware damage on M2M equipment, and M2M equipment should be able to regularly update antivirus software.

Data transmission is prevented from reaching the end of service, so that the attacker is able to obtain user data, signalling data or control data though physical theft or online listening to achieve the aim for unauthorized access. To obtain the secure communication between heterogeneous network systems, we can use the existing technology on embedded chip, which provides a singularly security architecture that operates as a security service to any application. It is uniquely serialized during manufacturing. It is easy to implement and there are no certificate servers to deploy, configure or maintain [31].

Privacy is defined in the area such as [32]:

  i. Storage
 ii. Processing
iii. Communication
 iv. Device

From security point of view, attributes mentioned below are of prime importance [33].

- Reliability refers to the fact that the service can be continued in spite of the system becoming vulnerable.
- For catastrophic systems, no security consequences are available there in nature.
- Maintainability stands for the ability of the normal system to undergo repairs and evolutions.
- Availability refers to the fact that data and systems can be accessed by authorized persons within an appropriate period of time.
- Integrity means the data or/and programs have not been modified or destroyed accidentally (e.g. transmission errors) or with malicious intent (e.g. sabotage).

Confidentiality describes the state in which sensitive information is not disclosed to unauthorized recipients. Table 2 shows existing work done in M2M security with parameters like message confidentiality, technology used, data authentication and solution for Man in the Middle Attack (MITM). A loss of confidentiality occurs when the contents of a communication or information

Table 2   M2M security state of the art

Ref. No	Existing Research	Data Integrity/ Authentication	Technology Used	Solution for Man in Middle Attack	Message Confidentiality
[1]	Internet of Things-New Security and privacy challenge	✓	Privacy enhancing technology (PET) like VPN, TLS, Onion routing	X	X
[2]	Cyber security for home user: a new way of protection through awareness enforcement	✓	E-Awareness Model used	X	✓
[3]	On the feature and challenges of security and privacy in distributed Internet of thing	✓	Cryptography algorithm	✓	X
[4]	The cyber threat landscape: challenges and future research direction	✓	Public private partner-ship(PPP)	X	✓
[5]	A framework for security quantification of networked machine	✓	Markov process tool used	X	✓

[6]	The evolution of M2M in IoT	✓	Store and forward, AAA services	X	X
[7]	Advancing M2M Communications Management: A Cloud-based System for Cellular Traffic Analysis	✓	Traffic management and fault tolerance technique	X	X
[8]	Large Scale Cyber-Physical Systems: Distributed Actuation, In-Network Processing and Machine-to-Machine Communications	✓	localized cooperative access, stabilization algorithm	X	X
[9]	Wireless Sensor Networking, Automation Technologies and Machine to Machine Developments on the Path to the Internet of Things	✓	Symmetric Algorithm, Hash Algorithm	X	✓

are leaked. Table 3 indicates currently available Security Provisions for IoT and M2M Communications.

5 Summary and Outlook

Next generation mobile communication will witness integration of various wired and wireless communication networks or services. IoT and M2M are going to hold maximum portion of all these wired and wireless communication networks because it has touched almost every possible field of communication. There exist security protocols for IoT and M2M communication but still new cyber-attacks are emerging every day. So cyber security solutions should

Table 3   State of the art security provisions for IoT and M2M communications

Entities	Protective Countermeasures
Consumers and end users	Guidance Provision for best practices to protect personal data and avoid problems in the mobile environment. Make the loading of applications and software permissions more intuitive and easier to understand.
Devices	Anti-malware and Anti- spam settings, Strong Authentication and Secure device connectivity.
Network-based security policies	Network operators provide numerous tools, guides and support to consumers, enterprise managers and end users to enable them to protect their information.
Authentication and controls for devices and users	Authorized access to Information Storage on mobile device system.
Cloud, networks and services	Aspects of these security solutions include consumer and enterprise applications, features for secure storage and virtual solutions and backup and disaster recovery
Security policies and risk management	Enhancements to security policies and risk management protocols; covering definitions and documentation; ongoing scans of the threat environment; and security assessments.
Monitoring and vulnerability scans	Automated periodic real time assessment of threats
Monitoring malware and cyber-threat profiles	From the cloud to Internet gateways, network servers and devices.
Wearable smart devices (watches, glasses)	Encryption, Authentication techniques
Smart Meters (information transport to utility provider)	Encryption
Home automation for convenience and protection	Encryption and VPN
Retail Near Field Communication (NFC)	Encryption, Malware security, Access controls

also grow accordingly and this is going to remain as a continuous process. Role based access control mechanisms are going to play vital roles in the robustness of the cyber security solution development for these services. Trust level based authentication mechanisms may result in the robust and secure communication.

References

[1] Georgios Tselentis, "Towards the Future Internet: A European Research Perspective", IOS Press, Netherlands, 2009.

[2] CTIA IoT White Paper, "Mobile Cybersecurity and the Internet of Things – Empowering M2M Communication", May 2014.

[3] Luigi Atzori, Antonio Iera, Giacomo Morabito, "The Internet of Things: A Survey", Elsevier 2010.

[4] Elgar Fleisch, "What is the Internet of Things? – An Economic Perspective", Auto-ID Labs White Paper WP-BIZAPP-053, January 2010.

[5] Sudha Nagesh, "Roll of Data Mining in Cyber Security" Journal of Exclusive Management Science, Vol. 2 Issue 5, pp. 2277–5684, May 2013.

[6] A. Q. Ansari, Tapasya Patki, A. B. Patki, V. Kumar, "Integrating Fuzzy Logic and Data Mining: Impact on Cyber Security", Fourth International Conference on Fuzzy Systems and Knowledge Discovery, FSKD 2007.

[7] Chen Hongsong, "Security and Trust Research in M2M System", IEEE International Conference on Vehicular Electronics and Safety (ICVES), 2011.

[8] Fibocom Perfect Wireless Experience, "The Coming Boom in Smart Buildings", http://www.fibocom.com/news/2-4-4-2-44.html

[9] Smart Grid News, "Smart grid advice from SMUD (Home gateways? Really?) http://www.smartgridnews.com/story/smart-grid-advice-smud-home-gateways-really/2014-07-21

[10] Nano Markets, "Smart-Grid Sensor Market Steadily Climbing to $39 Billion by 2019", http://nanomarkets.net/news/article/smart-grid-sensor-market-steadily-climbing-to-39-billion-by-2019-according

[11] M2M Now, Sep 12 2014, "Electric scooter with smartphone connection is launched in Europe by Terra Motors". http://news.ne2ne.com/articles/951843/electric-scooter-with-smartphone-connection-is-lau/

[12] Fibocom Perfect Wireless Experience, "Smart connected homes driving IoT". http://www.fibocom.com/news/2-4-4-2-39.html

[13] Machina Research, m2m now- Latest Machine to Machine Industry News, Connected Car Report http://www.connectedcar.org.uk/machinaltem2mnow/4586932250

[14] Machina Research, m2m now- Latest Machine to Machine Industry News http://www.m2mnow.biz/2013/10/25/16033-technology-migration-strategies-of-us-carriers-will-trigger-a-new-era-of-lte-for-the-m2m-industry/

[15] Machina Research, m2m now- Latest Machine to Machine Industry News http://www.m2mnow.biz/2013/11/26/16748-the-connected-car-a-usd282-billion-opportunity-but-who-pays/

[16] Machina Research, m2m now- Latest Machine to Machine Industry News http://www.m2mnow.biz/2013/12/13/17392-the-rise-of-m2miot-platforms-highlights-new-commercial-dynamics-and-new-challenges/

[17] Telematics Business Arena | Avlgps M2m Lbs | Fleet Management, Tracking, Rastreo, Gestion De Flotas Https://Www.Linkedin.Com/Groups/Berg-Insight-Says-28-Million-1878321.S.204554388

[18] m2m now- latest machine to machine industry news http://www.m2mnow.biz/2013/06/17/12936-m2m-the-focus-is-still-on-people/

[19] ABI Research, Technology Market Intelligence. https://www.abiresearch.com/press/v2v-penetration-in-new-vehicles-to-reach-62-by-202

[20] Yi Cheng, Mats Naslund, Goran Selander, and Eva Fogelstrom, "Privacy in Machine-to-Machine Communications", IEEE International Conference on Communication Systems (ICCS), 2012.

[21] A. Q. Ansari, Tapasya Patki, A. B. Patki, V. Kumar, "Integrating Fuzzy Logic and Data Mining: Impact on Cyber Security", Fourth International Conference on Fuzzy Systems and Knowledge Discovery, FSKD 2007.

[22] Radrigo Roman, Jianiying Zhou, "On the feature and challenges of security and privacy in distributed Internet of Things", Elsevier Journal on Computer Networks, Volume 57, Issue 10, Pages 2266–2279, July 2013.

[23] Danny Palmer, Computing News Security Threats and Risks, 2014. http://www.computing.co.uk/ctg/news/2323661/cyber-attack-launched-through-fridge-as-internet-of-things-vulnerabilities-become-apparent

[24] Kaoru Hayashi, Computing News Security Threats and Risks, 2013. http://www.computing.co.uk/ctg/news/2309972/new-internet-of-things-worm-discovered

[25] Orlando Debruce, Proofpoint U.S. http://www.computing.co.uk/ctg/news/2309972/new-internet-of-things-worm-discovered

[26] M2M – Latest Machine to Machine Industry News http://www.m2mnow.biz/2014/09/10/25004-9-meals-anarchy-take-cybersecurity-threat-iot-seriously-says-beecham-report/

[27] Manu Namboodiri, "Thoughts on M2M and IoT security and privacy for 2015" Blog on Thoughts about M2M and Internet of Things as well as related worlds from the M2Mi team.

[28] Michael Huth N. Asokan, Srdjan Capkun Ivan Flechais, Lizzie Coles-Kemp (Eds.), "Trust and Trustworthy Computing", 6th International Conference, TRUST 2013, London, UK, June 2013 Proceedings, Springer.

[29] Vangelis Gazis, Konstantinos Sasloglou, Nikolaos Frangiadakis, and Panayotis Kikiras, "Wireless Sensor Networking, Automation Technologies and Machine to Machine Developments on the Path to the Internet of Things," 16th Panhellenic Conference on Informatics, 2012.

[30] Ivan Stojmenovic, "Large Scale Cyber-Physical Systems: Distributed Actuation, In-Network Processing and Machine-to-Machine Communications", Mediterranean Conference on Embedded Computing, 2013.

[31] Giacomo Ghidini, Stephen P. Emmons, Farhad A. Kamangar, and Jeffrey O. Smith, "Advancing M2M Communications Management: A Cloud-based System for Cellular Traffic Analysis", 15th International IEEE Symposium on A World of Wireless, Mobile and Multimedia Networks (WoWMoM), 2014.

[32] Min Chen, Jiafu Wan, "A Survey of Recent Developments in Home M2M Networks", IEEE Communications Surveys and Tutorials, Volume: 16, Issue: 1, 2014.

[33] Hui Wang, Suman Roy, Amitabha Das, and Sanjoy Paul, "A Framework for Security Quantification of Networked Machines", Second International IEEE Conference on Communication Systems and Networks (COMSNETS), 2010.

Biographies

V. M. Rohokale received her B.E. degree in Electronics Engineering in 1997 from Pune University, Maharashtra, India. She received her Masters degree in Electronics in 2007 from Shivaji University, Kolhapur, Maharashtra, India. She received her PhD degree from CTIF, Aalborg University, Denmark under the guidance of Prof. Ramjee Prasad. She is presently working as Dean, R and D at SKN Sinhgad Institute of Technology and Sciences (SKN-SITS), Lonavala, Maharashtra, India. Her research interests include Cooperative Wireless Communications, AdHoc and Cognitive Networks, Physical Layer Security, Information Theoretic security and its Applications, Cyber Security, etc.

R. Prasad is currently the Director of the Center for TeleInFrastruktur (CTIF) at Aalborg University, Denmark and Professor, Wireless Information Multimedia ommunication Chair.

Ramjee Prasad is the Founder Chairman of the Global ICT Standardisation Forum for India (GISFI:www.gisfi.org) established in 2009. GISFI has the purpose of increasing of the collaboration between European, Indian, Japanese, North-American and other worldwide standardization activities in the area of Information and Communication Technology (ICT) and related application areas. He was the Founder Chairman of the HERMES Partnership – a network of leading independent European research centres established in 1997, of which he is now the Honorary Chair. He is a Fellow of the Institute of Electrical and Electronic Engineers (IEEE), USA, the Institution of Electronics and Telecommunications Engineers (IETE), India, the Institution of Engineering and Technology (IET), UK, Wireless World Research Forum (WWRF) and a member of the Netherlands Electronics and Radio Society (NERG), and the Danish Engineering Society (IDA). He is also a Knight ("Ridder") of the Order of Dannebrog (2010), a distinguished award by the Queen of Denmark. He has received several international award, the latest being 2014 IEEE AESS Outstanding Organizational Leadership Award for: "Organizational Leadership in developing and globalizing the CTIF (Center for TeleInFrastruktur) Research Network". He is the founding editor-in-chief of the Springer International Journal on Wireless Personal Communications. He is a member of the editorial board of other renowned international journals including those of River Publishers. Ramjee Prasad is a member of the Steering committees of many renowned annual international conferences, e.g., Wireless Personal Multimedia Communications Symposium (WPMC); Wireless VITAE and Global Wireless Summit (GWS). He has published more than 30 books, 900 plus journals and conferences publications, more than 15 patents, a sizeable amount of graduated PhD students (over 90) and an even larger number of graduated M.Sc. students (over 200). Several of his students are today worldwide telecommunication leaders themselves.

How to Use Garbling for Privacy Preserving Electronic Surveillance Services

Tommi Meskanen[1], Valtteri Niemi[1] and Noora Nieminen[1,2]

[1] *Department of Mathematics and Statistics, University of Turku,*
20014 Turun yliopisto, FINLAND
[2] *Turku Centre for Computer Science (TUCS), FINLAND*
Corresponding Authors: {tommes; pevani; nmniem}@utu.fi

Received 15 September 2014; Accepted 17 April 2015;
Publication 22 May 2015

Abstract

Various applications following the Internet of Things (IoT) paradigm have become a part of our everyday lives. Therefore, designing mechanisms for security, trust and privacy for this context is important. As one example, applications related to electronic surveillance and monitoring have serious issues related to privacy. Research is needed on how to design privacy preserving surveillance system consisting of networked devices. One way to implement privacy preserving electronic surveillance is to use tools for multiparty computations. In this paper, we present an innovative way of using garbling, a powerful cryptographic primitive for secure multiparty computation, to achieve privacy preserving electronic surveillance. We illustrate the power of garbling in a context of a typical surveillance scenario. We discuss the different security measures related to garbling as well as efficiency of garbling schemes. Furthermore, we suggest further scenarios in which garbling can be used to achieve privacy preservation.

Keywords: Internet of Things, privacy, electronic surveillance, garbling schemes.

Journal of Cyber Security, Vol. 4, 41–64.
doi: 10.13052/jcsm2245-1439.413

1 Introduction

Nowadays, we are surrounded by an increasing variety of *things* or *objects* that are connected with each other and accessible through the Internet. This trend is a consequence of a novel paradigm, Internet of Things (IoT), in which the devices form a network configured to reach goals common to all devices. The paradigm itself has gained increasing interest after the introduction of technologies that enable computing-like devices to share their states through the common network. These technologies include *Radio-frequency identification tags* (RFID) [10], *Near-field communication* (NFC) techniques and *Wireless sensor and actuator network* (WSAN) [27]. As an example of a network of devices trying to reach a common goal, consider an anti-theft system with motion detecting sensors. The sensors located differently interact with each other in order to detect unauthorized motion and prevent intruders. Many other applications of IoT can be found in [3, 27]. Some of the applications mentioned in [3] have been collected into Figure 1.

1.1 Related Research on Security

The technological advances alone are not sufficient to guarantee success for IoT-based solutions – the security of the technology is an important aspect as well. There are a variety of security threats related to IoT, as Roman et al. show in [26]: The threats are targeted at infrastructure, protocol and network security, data and privacy, identity management, trust and governance as well as at fault tolerance. For example, current Internet protocols may not meet the

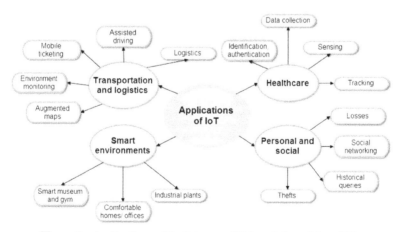

Figure 1   Applications of the Internet of Things (adapted from [3]).

security requirements of IoT, especially in the IP-based IoT as Heer et al. show in [18]. The physical security of IoT devices, e.g. tamper-resistance, is also an important aspect. An overview of different threats and possible solutions can be found in [27], whereas a more detailed threat analysis of RFID can be found in [10] and analyses of NFC from [17].

Several security threats are also identified by Kozlov et al. in [20]: there are numerous scenarios which endanger the security, trust or privacy of the IoT and these issues must be taken into account when considering legislation related to the Internet of Things. According to Weber [28], the IoT technology used by private enterprises must have resilience to attacks, authenticate the retrieved address and object information, have an access control and ensure client privacy. The privacy concerns are notable in situations in which the actions of individuals are monitored in a privacy-sensitive context. For example, a failed implementation of IoT related technology in a supermarket may violate the client privacy by enabling "the mining of medical data, invasive targeted advertising, and loss of autonomy through marketing profiles or personal affect monitoring" [29]. However, innovative ways of deploying privacy preserving IoT in privacy sensitive environments successfully are also possible: Abie et al. consider risk-based adaptive security framework for IoT in eHealth in [2]. More generally, techniques to achieve privacy preserving IoT applications have been considered widely. For example, privacy preserving electronic surveillance [11, 24] and even privacy preserving data mining [8] are possible by using a set of powerful cryptographic methods, called *secure multiparty computation* (SMPC).

1.2 Our Contributions

The solutions to achieve SMPC include a variety of protocols, e.g. *oblivious transfer* [25], *secure sum protocols* [8] and *garbled circuits* [30]. In this paper, we consider a way of achieving privacy preserving IoT applications by applying SMPC protocols. More specifically, we introduce a new way to realize privacy preserving electronic surveillance.We present a new tool in this context, *garbling*, which enables private computation on encrypted data.

The paper is organized as follows. We demonstrate the power of garbling in an example scenario presented in Section 2. In Section 3, we describe the realization of the privacy preserving electronic surveillance. In Section 4, we analyze the novel application of garbling in more details by considering its efficiency and what kinds of security goals are achieved by this technique. Section 5 concludes the paper and proposes directions for future research related to privacy preserving electronic surveillance.

2 Problem Setting: Privacy Preserving Electronic Surveillance System

Electronic surveillance is an application where privacy is a central concern. Many cryptographic tools have been proposed to ensure at least some level of privacy. In this paper, we present an innovative way of achieving privacy by using garbling in the context of electronic surveillance. Let us consider the following scenario as an example.

The client in this scenario is an elderly person living alone who wants to use the security service provided by a security company. The security company bases its service on an electronic surveillance system consisting of *Closed-Circuit Televisions* (CCTV) and various sensors (for example, motion detectors and/or sensors measuring the activity of the client). The security company collects data obtained by the system for further analysis. The analysis process contains tools for *data mining*, *pattern recognition* and *machine learning* – the intelligent surveillance system is supervised to react correctly on different situations. In certain situations (for instance, when the ongoing event seems to differ significantly from the usual course of events), the system evokes an alarm. The alarm together with an assessment of the situation enables the security company to react appropriately to the situation (e.g. call police/ambulance, send a guard from the company or just notify the client). The security company has outsourced its data center services into *a cloud* managed by a third-party company. The data from the surveillance system is stored and analyzed entirely in the cloud environment.

The main concern in this scenario is how the privacy for the client is managed. First obvious requirement is that anyone beyond the client, the security company and the cloud should not learn the contents of the data collected by the surveillance system. This requirement can be reached by simply encrypting the data on the client side and decrypting the data on the security company side. As a consequence, the security company and the cloud provider can follow everything that is going on at the client's home. A serious concern is that the third-party company managing the cloud can learn something about the client that could be used for unwanted or even malicious purposes. Thus it is highly justified to hide the raw data also from the cloud whereas the security company needs the raw data to be able to react correctly in the alarming situations. A solution to this is to use two-party computation between the security company and the cloud. This would allow the cloud to analyze the surveillance data without allowing the cloud to learn the raw data or the analytics tools.

However, this solution is still problematic. The security company should monitor the surveillance data of numerous customers in real time while the analysis on cloud is ongoing. This is not desirable because of several reasons. From the company's perspective, real-time monitoring is inefficient – several employees are tied to follow the monitors and are demanded to be in alert readiness all the time, even though nothing alarming is happening. From the client's perspective, the all-time surveillance is distracting and feels privacy violating – the security company should be able to study the raw data only in alarming situations and not otherwise.

To summarize the above analysis of the scenario, the implementation of the privacy preserving electronic surveillance system should have the following properties.

Confidentiality: All the information related to the electronic surveillance is kept secret from parties excluding the client, the security company and the third-party cloud. The third-party cloud performs the analysis on encrypted data. The cloud retrieves the encrypted surveillance data from the client and the encrypted surveillance data from the security company. The cloud is not allowed to find out the unencrypted surveillance data (the data is privacy-sensitive) or the analytics tool (the tool may be intellectual property of the security company). Depending on the contract between the security company, the client and the cloud, the final analysis result can either be concealed from the cloud or can be revealed to the cloud. These alternatives are discussed in more detail in Section 4. The security company is not allowed to retrieve the unencrypted surveillance data unless the analysis result yields an alarm. The client is not allowed to learn the implementation details of the analytics tool (the tool may be intellectual property of the security company).

Integrity: We may assume that the client is honest and therefore the surveillance data is authentic. Cloud can be honest, semi-honest or even malicious – a garbling scheme achieving certain level of security (explained in Section 4) guarantees that the analysis result is also authentic. Additionally, integrity of data in transit is protected, e.g. by using message authentication codes.

Entity authentication: The cloud does not need to authenticate itself, since all the data it processes is encrypted (in the case the cloud is not allowed to find out the analysis result). The security company and the client authenticate themselves when the system is first configured. After the authentication, we assume that the channel between the client and the security company is confidential and authentic.

Access control: The security company is able to retrieve the unencrypted surveillance data only in the case in which the final analysis result yields an alarm. This requires that the analytics tools must not reveal the surveillance data. To ascertain this, the client and the security company use a trusted auditor that verifies appropriateness of the analytics tool (the analytics tool does not leak surveillance data in the final analysis report). We have described the different solutions for accessing the unencrypted surveillance data in Section 3.1.

Authorization: Access control, entity authentication and other security measures naturally require that access to various resources is properly authorized. For example, the client has to authorize the security company to have access to raw data in case of alarm and the security company has to provide authorization for the cloud provider in order to receive garbled data from the client.

Non-repudiation: We assume that the garbling protocol will achieve authenticity. This guarantees that the cloud cannot forge the garbled evaluation, and that the encrypted analysis result is authentic. We assume that the surveillance data is authentic. We also assume that the channel between the client and the security company is confidential and authentic. We also assume that the cloud service provider and the security company are not in the conspiracy against the client. Then log data collected by all parties can be used for non-repudiation purposes, see also discussion about logs in Section 3.2.

Availability: The system is naturally based on the assumption that raw data will be available for the security company in alarming situations. Related to this, there are threats purely concerning implementation. For example, burglars may cut off sending of data to the cloud. Also, the surveillance data stream may be interrupted on client side. As an example, robbers may break the CCTV equipment and sensors or the client may throw a towel on top of the surveillance camera etc.

   To achieve these properties, we need an additional tool that enables the cloud to evaluate the analytics algorithms on the surveillance data without learning anything about the algorithms or data. A tool that fulfills this requirement is garbling. The formal definition of garbling and different security aspects related to garbling can be found in Section 4. In the following section, we concentrate on how the surveillance system using garbling should be implemented.

3 Operating Model: How to Build Privacy Preservation in the Surveillance System

In this section, we describe the operating model that aims at a solution for the problem presented in the previous section: How can the security company provide privacy preserving electronic surveillance to an elderly person even when all the data services of the company have been outsourced to a third-party cloud provider.

Our solution is based on a cryptographic tool for secure multiparty computation, *garbling*. Garbling enables *secure and private function evaluation*. A user who does not have enough computing resources utilizes a possibly untrustworthy evaluator, such as cloud, to accomplish the evaluation of some function f on argument x. However, the user wants to keep both the function f and the argument x secret from the evaluator. The user and the cloud agree on using a garbling scheme that works as follows. First, the user garbles function f and its argument x and obtains garbled function F and garbled argument X. The user gives F and X to the evaluator who runs the garbled evaluation to get the garbled value $Y = Ev(F, X)$. Now, either the evaluator or the user ungarbles Y to get the final value y, which is equal to the result of the original evaluation $y = ev(f, x)$.

In the scenario presented in the previous section, the electronic surveillance system is designed under the IoT paradigm, making the computational resources of the system limited. This means that the surveillance data analysis must take place outside the surveillance system, for example at the data center of the security company. Since the data center services are outsourced, the analysis takes place in the cloud managed by a third-party company. The surveillance data from the client's home is privacy sensitive as are the analytics tools of the security company, so the three parties agree on using a garbling scheme. Figure 2 illustrates the scenario showing also how the garbling scheme is used by the different parties.

3.1 Responsibilities of the Different Parties

The surveillance data is garbled on the client side. In this way, neither the cloud nor the security company is able to access the raw data directly. The garbled surveillance data is sent to the cloud for analysis. The security company has garbled its analytics tools that act as the function to be evaluated on cloud. After receiving both the garbled data and the garbled analytics tools, the cloud runs the garbled evaluation getting the garbled final value. If the cloud is allowed to decrypt the garbled analysis result, then it decrypts the

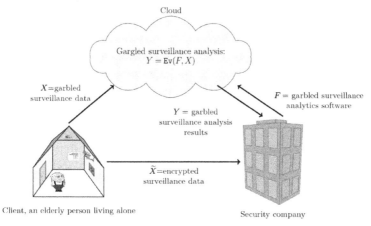

Figure 2   Scenario about electronic surveillance.

garbled value getting the final outcome of the analysis. This final outcome is sent to the security company for further investigation. However, letting the cloud learn the analysis results might not be convenient – it can violate the privacy in similar manner as the actual surveillance data. Thus, a more convenient way of implementation is that the cloud sends the garbled analysis outcome to the security company for further investigation. Now, the security company ungarbles the data received from the cloud. Based on the analysis outcome, the security company takes corresponding actions (e.g. by visiting the home or calling the police, an ambulance, a social worker, the person's relatives etc.).

A straight-forward way of reacting to the alarm situation for the security company is that a guard from the company visits the client for further inspection in spite of what has caused the alarm. Then, the security company does not have or even need an access to the raw data (in Figure 2 this means, that no encrypted surveillance data $\tilde{X}$ is provided to the security company). However, this is not a practical approach. The company should adaptively react to different alarms – for example a robbery should cause different reactions than the client staying suspiciously long in the shower.

To adaptively react to the various situations, the security company needs an access to the raw data. Granting the security company access to the raw data with no restrictions is not a satisfactory solution since it would violate the requirements set to the system: the raw data should be accessible for the security company only in alarming situations (see Confidentiality requirement in Section 2). One possible way of realizing the access control would be to encrypt the raw data twice, independently for the cloud and for the security company. The raw data is protected against the cloud by garbling the argument x. Garbling x is modeled by encrypting x using the encryption algorithm En together with encryption key e. Respectively, the algorithm De with the decryption key d is used to ungarble Y to final value y. The encryption against the security company utilizes an independent encryption algorithm $\widetilde{En}$ with key $\tilde{e}$. The decryption algorithm $\widetilde{De}$ with the decryption key $\tilde{d}$ is used to recover the raw data from the encrypted data $\widetilde{X}$ – this key is called *recovery key* to avoid confusion between the two keys d (which is needed for ungarbling) and $\tilde{d}$ (which is needed for recovering x from $\widetilde{X}$).

The surveillance data is collected in pieces and these pieces are then encrypted and sent to the cloud (X) and to the security company ($\widetilde{X}$) by the client. Data pieces may contain overlaps so that successful reconstruction of the course of events without gaps is possible. Since the cloud does not learn the keys (e, d), and hence learns nothing privacy – violating about the surveillance data x or the analysis result y, the same keys (e, d) may be used for many evaluations by the analytics tool.

The same does not hold for the keys $(\tilde{e}, \tilde{d})$ related to the encryption of surveillance data against the security company. If the same keys $(\tilde{e}, \tilde{d})$ were used, then the security company would-be able to follow all the surveillance data after accessing the keys $(\tilde{e}, \tilde{d})$ for the first time – even in the non – alarming situations. This clearly violates the privacy policy we have set to the system. Thus, a more sophisticated access control method is needed. We have identified the following two approaches to implement access to the recovery key $\tilde{d}$.

3.2 The First Approach

In this approach, the recovery key $\tilde{d}$ for recovering the encrypted surveillance data $\widetilde{X}$ is possessed by the security company. However, the key $\tilde{d}$ must be protected by an electronic seal because otherwise the company could decrypt all the surveillance data and not only the data related to alarms. The company is allowed to break the seal whenever the analysis yields an alarm. After breaking the seal, the company uses the recovery key to obtain the actual surveillance data consisting of the moments some time before and after the alarm.

The above approach requires a countermeasure to detect unauthorized access to the surveillance data. One possible way is to utilize event logging. Each of the three parties related to the surveillance are maintaining their own independent event logs. The independent logs contain information that can be derived from the activities of the different parties (for example, the company logs access to the raw data together with a synopsis of the analysis results). These three independent logs can in principle be compared to detect unauthorized or illegitimate access to the backup data. Of course, the different logs can be forged and the comparison does not work in the desired way in case there are conspiracies between the parties but solving conspiracy issues is not in the scope of this paper.

In this approach, the efficiency of the implementation depends on the efficiency of the used garbling scheme as well as the efficiency of the used independent encryption scheme $\mathcal{E} = (\widetilde{\text{KeyGen}}, \widetilde{\text{En}}, \widetilde{\text{De}})$. The efficiency of garbling schemes is discussed in more detail in Section 4.3. The efficiency of the encryption scheme $\mathcal{E}$ is due to the choice of the security company. For example, $\mathcal{E}$ may be AES-128.

3.3 The Second Approach

In this approach, the recovery key $\tilde{d}$ is in client's possession. Since the security company does not possess the decryption key $\tilde{d}$ of $\tilde{X}$, the company cannot monitor the data unless it is handed the decryption key. The company should be able to get the decryption key only in alarming situations. A straight-forward way of implementing the access control into the recovery key $\tilde{d}$ is to use timestamped key management (see [19] for further information). The security company can access the keys $\tilde{d}$ related to the raw data having certain timestamps that correspond to the time of the alarm detection as well as the data from some moments before and after the alarm detection. The client can later check which keys have been sent to the security company and, if needed, check the corresponding raw data.

There is also a more innovative way of implementing the access control into the recovery key $\tilde{d}$. Informally, our idea is to send the recovery key $\tilde{d}$ to the security company via the cloud in such a way that the cloud does not learn the recovery key. Moreover, the security company will receive the key only in the case that the final analysis results yield an alarm. Next, we explain in more details, how this functionality can be implemented.

For simplicity, let us assume that the final surveillance analysis result is either alarm or no alarm, i.e. $y \in \{\text{alarm, no alarm}\}$. Now, we want that the security company gets $\tilde{d}$ whenever $y = \text{alarm}$. This can be reached by attaching first the recovery key $\tilde{d}$ to the surveillance data, i.e. $x_m = (x, \tilde{d})$. This argument is then garbled and sent to the cloud, thus the cloud is not able to learn $\widetilde{d}$. The function f needs to be modified in order to be able to handle the new argument type. We define the modified function as follows

$$f_m(x_m) = \begin{cases} (y, \widetilde{d}) & \text{if } y = \text{alarm} \\ (y, \varepsilon) & \text{otherwise (where } \varepsilon \text{ is the empty string)} \end{cases}$$

The garbling scheme works in a similar manner as before. The cloud gets the garbled argument X_m from the client and the garbled function F_m from the security company. The cloud computes the garbled value Y_m and sends it to the security company. The security company ungarbles Y_m and gets $y_m = (y, \beta)$. Here, $\beta \in \{\widetilde{d}, \varepsilon\}$ depends on whether $y = \text{alarm}$ or not.

The modifications in x and f now give the required functionalities. First of all, the security company gets the recovery key $\widetilde{d}$ only in the case y yields an alarm. Secondly, sending the key via the cloud is not insecure – the key remains garbled during the whole garbled evaluation in similar manner as the argument and the function.

The efficiency of implementation using this approach depends on the efficiency of the used garbling scheme and the efficiency of the used encryption scheme $\mathcal{E} = \left(\text{KeyGen}, \widetilde{\text{En}}, \widetilde{\text{De}} \right)$. However, the function f_m and the argument x_m are more complex than in the first approach, since the argument x_m contains the recovery key $\widetilde{d}$ and the function f_m needs to process $\widetilde{d}$ somehow. This means, that the second approach is not as efficient as the first approach. On the other hand, the second approach provides better control over the use of the recovery key $\widetilde{d}$.

4 Implementation of the Surveillance Service

In this section, we describe garbling schemes in more details. We start by defining the concept after which we discuss the different security measures for garbling schemes. We also discuss which of the security concepts are ideal for the use in the context of privacy preserving electronic surveillance.

4.1 Building Blocks

As mentioned earlier, our main building block to construct a privacy preserving and cloud-assisted surveillance system is garbling. Garbling enables surveillance data to be analyzed on cloud environment without compromising the privacy of the client or revealing business secrets in the form of the analytics tool.

The surveillance analytics tool may contain algorithms e.g. for anomaly detection [7, 9] (to detect the abnormal situations among the normal situations), and for machine learning. Recently, a method for running machine learning algorithms on encrypted data has been proposed [16].

One possible way of teaching the analytics tool is the following. Before the surveillance starts, the company and the client may have collected data from normal situations. These labeled situations together with the data from the surveillance system act as the training data for the semi-supervised learning (see [31] for more information) algorithm that now helps in doing the final analysis together with the other algorithms. We do not concentrate on the exact implementation of the analytics tool as our focus is on the tools enabling the privacy preserving surveillance.

4.2 Formal Definitions for Garbling

Formally, a garbling scheme is a 6-tuple of algorithms, (KeyGen, Ga, En, De, Ev, ev). The last component of the tuple is the evaluation algorithm ev: an algorithm that computes the value of function f on argument, i.e. $y = f(x)$. In our scenario, the function f is the surveillance analytics tool and the argument x is the surveillance data. To hide the analytics tool and the surveillance data, both f and x are garbled. To do this, first key generation algorithm KeyGen is called to generate three keys $(g,\ e,\ d)$. The garbling algorithm Ga computes the garbled function $F = Ga(g, f)$ based on the function f and garbling key g. The encryption algorithm En computes the garbling $X = En(e, x)$ based on argument x and encryption key e. The garbled evaluation function (the garbled analytics tool) computes the garbled value $Y = Ev(F, X)$ (garbled analysis). Finally, the decryption algorithm De ungarbles Y and returns the final analysis result $y = De(d, Y)$ by using the decryption key d issued by the KeyGen algorithm. Note that the final analysis result must be the same despite of the method used for evaluation: the garbled evaluation must yield the same final analysis result as the actual evaluation, i.e. $ev(f, x) = y = De(d, Y) = De(d, Ev(F, X))$. The garbled evaluation process is illustrated in Figure 3. For further details, consult e.g. [22].

In the example scenario presented in this paper, a garbling scheme $G = $ (KeyGen, Ga, En, De, Ev, ev) is used as follows. The client in the scenario uses the algorithms KeyGen and En. The security company uses algorithm Ga. The cloud uses algorithm Ev. Depending on the case, either the cloud or the security company uses algorithm De. Figure 4 illustrates how the different algorithms are run by different parties in the example scenario.

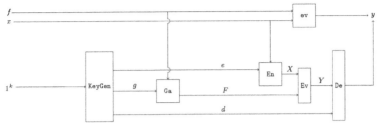

Figure 3 The components and the workings of a garbling scheme. The diagram is adapted from [22].

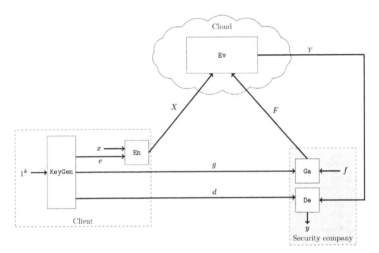

Figure 4 Garbling scheme in the surveillance scenario.

In the illustration, we assume that the channel between the client and the security company assures data integrity, authenticity and confidentiality. This is not assumed for the channel between the client/the security company and the cloud. We also present the situation in which the security company is the party ungarbling Y.

4.3 Security Considerations

In this section, we first introduce different security concepts for garbling schemes. Exact definitions for each concept can be found in literature [6, 5, 21–23]. Then, we analyze which of the security concepts meet the requirements for the privacy preserving electronic surveillance system proposed in the previous section.

Every security concept can be characterized by *security notion* and *level of reusability*. The security notion tells what kind of information about the function f and the argument x is allowed to be leaked. The notion *function and argument hiding* means that the garbling scheme is allowed to leak $f(x)$, but neither f nor x. The notion *function, argument and final value hiding* does not allow the garbling scheme to leak any of f, x or $f(x)$. The notion *matchability-only* does not allow the garbling scheme to leak f nor x, but when evaluating f on two different arguments x_1 and x_2, the garbling scheme is allowed to leak whether $f(x_1) = y_1 = y_2 = f(x_2)$. Note that the names of the notions differ from the ones used in literature. We use non-standard names to distinguish what we mean by privacy in the example scenario and privacy related to garbling schemes. The notion *function and argument hiding* corresponds the notion *privacy* in [5, 23]. The notion *function, argument and final value hiding* corresponds the notion *obliviousness* in [5, 23].

The security notions described above deal with secrecy. The *authenticity* property can also be formalized for garbling schemes. Authenticity guarantees that an adversary is unable to create a garbled value Y from a garbled function F and its garbled argument X such that $Y \neq F(X)$ but which will be considered authentic. For a formal definition of authenticity for garbling schemes, consult [6]. The authenticity of garbling schemes is needed to fulfill requirement 2, demanding that the final analysis result must be authentic. Thus, the garbling scheme used in the example scenario must achieve authenticity in the sense explained in [6].

Another characteristic of a garbling scheme is *the level of reusability*. The level of reusability tells how many times the same garbled function can be securely used for different arguments. The first reusability level enables only one-time use of the same garbled function [5, 6, 21] whereas higher levels of reusability enable several or even arbitrary reuse of the garbled function [15, 23].

Let us first recall Confidentiality, Integrity, Entity authentication, Access control, Authorization, Non-repudiation and Availability requirements presented in Section 2. The Confidentiality requirement says that the unencrypted surveillance data must be kept secret from third parties, including the cloud. Moreover, the analytics tool must be hidden from third-party cloud. From the garbling point-of-view, this means that the garbling scheme should be at least function and argument hiding. The Confidentiality requirement also tells that hiding the final analysis result depends on the contract between the client, security company and the cloud.

We have identified three possible configurations of the surveillance system all of which set different security requirements to the garbling scheme in use. In all three cases, the surveillance data as well as the surveillance data analytics tools are kept secret from the cloud. The differences in the configurations are related to the final analysis results: are the analysis results kept totally, partially or not at all secret from the cloud. Let us next provide more details of these three different configurations.

Case 1: The cloud is allowed to learn nothing about the resulting analysis. This means that the surveillance data, the analytics tool and the final analysis result are all hidden from the cloud. From the garbling point-of-view, this is the same as hiding the function, the argument and the final value. This is desirable, because the third-party company may use the information about the analysis for its own purposes that might be unwanted by the client. Thus, the garbling scheme must leak none of f (the analytics tool), x (the surveillance data) or $y = \text{ev}(f, x)$ (the analysis result) to the cloud. A garbling scheme that is function, argument and final value hiding meets these requirements.

Case 2: The cloud is allowed to learn indirect information about the analysis result but possibly not the actual content of the final analysis. From the garbling point-of-view, this is the same as hiding the function, the argument and the final value but leaking some information about the final value. One justification for this weaker privacy requirement is the following: the cloud service provider may anyway be able to find out the actions of the security company related to certain garbled analysis results. For example, the cloud has found that garbled analysis result Y yields a call to the police. When the same garbled analysis result Y is found later again on the cloud, the cloud service provider is able to predict the reaction of the security company – the security company will probably call to the police.

Thus, we could require that the cloud service provider cannot find out f, x or y but it is able to find out whether the certain Y yields similar actions as before. For the garbling scheme this means that the scheme should not leak f, x or y but it may leak whether $f(x_1) = y_1 = y_2 = f(x_2)$ when computing $\text{ev}(f_1, x_1)$ and $\text{ev}(f_2, x_2)$. This is exactly what a garbling scheme achieving matchability-only security provides.

Case 3: The cloud service provider is allowed to learn the final analysis result. From the garbling point-of-view, this means that the garbling scheme is allowed to leak the final value y whereas it must hide the function and the argument. However, the Confidentiality requirement tells that the cloud is allowed to learn nothing about the ungarbled surveillance data, meaning that the final analysis cannot contain parts of surveillance data. To assure this, the final value could be something else than a review of the surveillance data. It may also be *the type of alarm*, like no alarm, low urgency, medium urgency, high urgency etc. or simply alarm/no alarm. Now it may under some circumstances be acceptable to let the cloud provider to know the type of the alarm. This means the garbling scheme should hide f and x but it may leak y. This is exactly what a function and argument hiding garbling scheme provides. However, it is questionable whether the cloud should generally learn that there is an alarming situation at the client. This violates the requirement that the third – party cloud should learn nothing about the surveillance, not even the fact that the security company is being alarmed.

The above reasoning suggests that the garbling scheme should either be function, argument and final value hiding or achieve matchability-only. From the practical point of view, matchability-only is preferable as it has been shown in [21–23] that it is at least as easy to achieve matchability-only as to be function, argument and final value hiding. Moreover, for practical reasons one should be able to use the same garbled analytics tool for several garbled surveillance data entries. For the garbling scheme this converts to reusability. Thus we suggest that the applied garbling scheme should be reusable as well as achieve matchability-only and authenticity. This guarantees that the surveillance is privacy-preserving since the third parties do not learn the ungarbled surveillance data or the ungarbled analysis result.

4.4 Efficiency Considerations

The above concepts do not restrict the computation method for evaluating function f on argument x – the function f may represent models such as a circuit, a Turing machine or a random-access machine. Methods for garbling various computational models have been constructed. For example, there exist garbling schemes for circuits [30], Turing machines [14] and random-access machines [13].

The choice of the computation method affects the efficiency of the garbling scheme. Choosing circuits over Turing machines has at least two unfortunate consequences. The first consequence is related to the running time of circuits. The running time of a circuit is constant, implying that evaluating a circuit with any input takes the worst-case running time. This is not the case for Turing machines. Another unfortunate consequence is related to the size of the garbled function F. Turing machines outperform circuits also in this aspect: the size of garbled circuit is as large as the running time of the algorithm where as the size of the garbled Turing machine depends only on the description of the algorithm and not on the input value x. [14]

On the other hand, using circuits as the computation method has benefits over Turing machines when considering the costs of constructing the garbling scheme. Garbled circuits are known to have efficient constructions [4] where as such are not known for garbled Turing machines. Garbled Turing machines typically use fully-homomorphic encryption [12] as a building block which causes inefficiency in the construction.

Next we provide some numbers on the efficiency of garbling. The values have been collected from [4], in which three different garbling schemes have been experimented by using JustGarble (the source code is open-source and is available in [1]) system on an x86-64 Intel Core i7-970 processor clocked at 3.201 GHz with a 12MB L3 cache. The three garbling schemes are based on a function and value hiding garbling scheme Garble1 presented in [6]. All three presented garbling schemes are based on *dual-key cipher*. The difference in the three schemes is the different optimization techniques used to reduce the evaluation time. For more details, consult [4].

On a circuit having 15.5 million gates, of which 9.11 million gates are XOR gates, the most efficient garbling scheme *GaX* uses approximately 0.49 seconds to garble the circuit and 0.23 seconds to evaluate the garbled circuit [4]. This shows that the time for evaluating and garbling using *GaX* is efficient even on quite large circuits, meaning that even complex algorithms represented as circuits can be efficiently garbled and evaluated with *GaX*.

Using this measurement data presented in [4], we can estimate the efficiency of our solution for the example scenario as follows. Let us assume that the garbling scheme *GaX* is used. Let us further assume that the analytics tool is presented as a circuit having approximately 15.5 million gates. If the client sends surveillance data at rate of 0.5 kilobits/second then an analysis result is received by the security company approximately once per second. Achieving the low sending rate of 0.5 kilobits per second requires some pre-processing of the surveillance data on client side, e.g concerning the captured video stream where raw data is accumulated in much higher data rate. As a summary, these figures seem to be acceptable from practical point-of-view.

There are some known issues related to the use of scheme *GaX*. The garbled argument F is constructed at the same time as the keys (e, d), meaning that one party needs to possess both the function and the argument. We can solve this problem in two ways. First one is to let a trusted authority to run the garbling algorithm Ga with the garbling key g obtained from the client and the function f obtained from the security company. Another solution would be that the client and the security company use a secure multiparty computation protocol for computing $Ga(g, f)$ together in such a way that the security company does not learn g and the client does not learn f. Unfortunately, both solutions add to the complexity of the system and increase the time to garble.

Another problem in using any of the garbling schemes from [4] is that these garbling schemes are not reusable. This means that every time a new surveillance data entry is ready to be processed, both the surveillance data entry and the analytics tool need to be garbled. This implies that big amount of the total computation happens in the client side, and therefore the benefit of using cloud is questionable. In garbling schemes supporting reusable garbled circuits, the analytics tool represented as a circuit is garbled only once which would increase the overall efficiency of the garbling scheme. In our scenario, this would correspond to a situation where most of the computation load can be moved to the cloud. Unfortunately, no efficient constructions for reusable garbled circuits are known.

5 Discussion

Applications following the Internet of Things paradigm have increased rapidly, even to the extent that the security, trust and privacy related to the applications have not been able to keep up with the progress. Especially, privacy preservation seems to be one of the hottest topics related to the Internet of Things. One application that is regarded as privacy violating is electronic surveillance, at least in the private premises such as homes. On the other hand, there is a need to monitor private homes in order to track emergencies and protect the customers from various threats. We are confronting the challenging task of creating a privacy preserving electronic surveillance system.

In this paper, we have presented a novel way of using garbling schemes to achieve privacy preservation in electronic surveillance. We illustrated the power of garbling with an example scenario. An elderly person living alone is subscribing to a security service that includes electronic surveillance. The surveillance data is analyzed by a security company that has outsourced its data services onto a third-party cloud. Garbling allows the private analysis of the surveillance data on cloud – the cloud learns neither the surveillance data nor the analytics tool.

The example scenario is not the only possible application for garbling. As another related example, a monitoring system can be installed in the homes of people using the services for *assisted living*. The party monitoring the data should not learn the habits of the person using the system beyond the situations in which the person needs help. In this scenario, the security company may provide the monitoring services to the company providing the services for assisted living. This makes the privacy preservation even more complex task.

A variant of our example scenario presented in this paper is that the security company does not use third-party cloud services and instead does all the analysis itself. The operation of the parties present in this variant scenario resembles the operation of the same parties in the example scenario. However, there is one fundamental difference: the party possessing the analytics algorithms is the same as the party evaluating the analytics algorithms. This means that the security company first garbles the analytics algorithm as before, but then evaluates the garbled analytics algorithms on the garbled data received from the client.

In this case, the garbled evaluation might directly give the final analysis result y as hiding the result from the security company itself is useless. Moreover, the garbling of the analytics tool is not essential since the same party (the security company) both possesses the analytics tools and is responsible for the evaluation. Hence, the variant scenario is more efficient than the original scenario. However, moving the computation load from the cloud to the security company requires that the computational resources on the security company side should increase.

To conclude, we have found a novel solution to provide privacy preservation in an electronic surveillance system utilizing the Internet of Things paradigm. The biggest advantage in our solution is that garbling provides flexibility in the system. The surveillance analytics tool can be almost anything, from comparisons to complex machine learning algorithms. Moreover, the function f can be changed without need to reconfigure the whole system, easing the system maintenance.

The biggest obstacles for implementing the described system we have described is related to the implementation of efficient garbling schemes. There exist efficient garbling schemes (see Section 4.3) that support one-time use of the garbling scheme. But regarbling the analytics tool again for every surveillance data entry is not optimal from practical point-of-view. Reusable garbled circuits would solve this problem – however efficient garbling schemes supporting reusable garbled circuits are not known.

Future research may solve the problem of efficient reusable garbled circuits. Moreover, exploring further targets for innovative use of garbling in the context of IoT is important. An interesting target would be larger and more complex systems having more parties, for example a scenario where a person uses services for assisted living. In this scenario, we would have four parties – the client, the assisted living service provider, the security service provider (providing the devices for monitoring the client) and the third-party cloud service provider.

Acknowledgments

Authors would like to thank anonymous referees for valuable suggestions that have improved the paper quality significantly. This work was supported by the Academy of Finland project "Cloud Security Services" which is greatly appreciated.

References

[1] JustGarble. http://cseweb.ucsd.edu/groups/justgarble/. Accessed: 2014-10-13.

[2] H. Abie and I. Balasingham. Risk-based Adaptive Security for Smart IoT in eHealth. In *Proceedings of the 7th International Conference on Body Area Networks,* BodyNets'12, pages 269–275, ICST, Brussels, Belgium, Belgium, 2012. ICST (Institute for Computer Sciences, Social-Informatics and Telecommunications Engineering).

[3] L. Atzori, A. Iera, and G. Morabito. The Internet of Things: A Survey. *Computer Networks,* 54(15): 2787–2805, 2010.

[4] M. Bellare, V. T. Hoang, S. Keelveedhi, and P. Rogaway. Efficient garbling from a fixed-key blockcipher. In *Proc. of Symposium on Security and Privacy 2013,* pages 478–492. IEEE, 2013.

[5] M. Bellare, V. T. Hoang, and P. Rogaway. Adaptively secure garbling scheme with applications to one-time programs and secure outsourcing. In *Proc. of Asiacrypt 2012,* volume 7685 of LNCS, pages 134–153. Springer, 2012.

[6] M. Bellare, V. T. Hoang, and P. Rogaway. Foundations of Garbled Circuits. In *Proc. of ACM Computer and Communications Security (CCS'12),* pages 784–796. ACM, 2012.

[7] V. Chandola, A. Banerjee, and V. Kumar. Anomaly Detection: A Survey. *ACM Comput. Surv.,* 41(3): 15: 1–15: 58, July 2009.

[8] C. Clifton, M. Kantarcioglu, J. Vaidya, X. Lin, and M. Y. Zhu. Tools for Privacy Preserving Distributed Data Mining. *SIGKDD Explor. Newsl.,* 4(2): 28–34, Dec. 2002.

[9] T. Dunning and E. Friedman. *Practical Machine Learning: A New Look at Anomaly Detection*. O'Reilly Media, 2014.

[10] S. Evdokimov, B. Fabian, O. Günther, L. Ivantysynova, and H. Ziekow. RFID and the Internet of Things: Technology, Applications, and Security Challenges. *Foundations and Trends@in Technology, Information and Operations Management*, 4(2):105–185, 2011.

[11] K. B. Frikken and M. J. Atallah. Privacy Preserving Electronic Surveillance. In *Proceedings of the 2003 ACM Workshop on Privacy in the Electronic Society*, WPES '03, pages 45–52, New York, NY, USA, 2003. ACM.

[12] C. Gentry. *A Fully Homomorphic Encryption Scheme*. PhD thesis, Stanford University, 2009. crypto. stanford. edu/craig.

[13] C. Gentry, S. Halevi, S. Lu, R. Ostrovsky, M. Raykova, and D. Wichs. Garbled RAM Revisited. In *Proc. of 33rd Eurocrypt*, volume 8441 of LNCS, pages 405–422, 2014.

[14] S. Goldwasser, Y. Kalai, R. Popa, V. Vaikuntanathan, and N. Zeldovich. How to Run Turing Machines on Encrypted Data. In *Proc. of 33rd CRYPTO*, volume 8043 of LNCS, pages 536–553, 2013.

[15] S. Goldwasser, Y. Kalai, R. A. Popa, V. Vaikuntanathan, and N. Zeldovich. Reusable Garbled Circuits and Succinct Functional Encryption. In *Proc. of the 45th STOC*, pages 555–564. ACM, 2013.

[16] T. Graepel, K. Lauter, and M. Naehrig. ML Confidential: Machine Learning on Encrypted Data. In *International Conference on Information Security and Cryptology – ICISC 2012, Lecture Notes in Computer Science, to appear*. Springer Verlag, December 2012.

[17] E. Haselsteiner and K. Breitfuß. Security in near field communication (NFC). In *Workshop on RFID security*, pages 12–14, 2006.

[18] T. Heer, O. Garcia-Morchon, R. Hummen, S. L. Keoh, S. S. Kumar, and K. Wehrle. Security Challenges in the IP-based Internet of Things. *Wirel. Pers. Commun.*, 61(3): 527–542, 2011.

[19] A. V. D. M. Kayem, S. G. Akl, and P. Martin. Timestamped Key Management. In *Adaptive Cryptographic Access Control*, volume 48 of *Advances in Information Security*, pages 61–74. Springer US, 2010.

[20] D. Kozlov, J. Veijalainen, and Y. Ali. Security and Privacy Threats in IoT Architectures. In *Proceedings of the 7th International Conference on Body Area Networks*, BodyNets'12, pages 256–262, ICST, Brussels, Belgium, Belgium, 2012. ICST (Institute for Computer Sciences, Social-Informatics and Telecommunications Engineering).

[21] T. Meskanen, V. Niemi, and N. Nieminen. Classes of Garbled Schemes. *Infocommunications Journal*, V(3): 8–16, 2013.

[22] T. Meskanen, V. Niemi, and N. Nieminen. Garbling in Reverse Order. In *The 13th IEEE International Conference on Trust, Security and Privacy in Computing and Communications (IEEE TrustCom-14)*, 2014.

[23] T. Meskanen, V. Niemi, and N. Nieminen. Hierarchy for Classes of Garbling Schemes. In *Proc. of Central European Conference on Cryptology (CECC'14)*, 2014.

[24] V. Oleshchuk. Internet of things and privacy preserving technologies. In *1st International Conference on Wireless Communication, Vehicular Technology, Information Theory and Aerospace Electronic Systems Technology, 2009. Wireless VITAE 2009.*, pages 336–340, 2009.

[25] M. O. Rabin. How to Exchange Secrets with Oblivious Transfer. Technical report tr-81, Aiken Computation Lab, Harvard University, 1981.

[26] R. Roman, P. Najera, and J. Lopez. Securing the Internet of Things. *Computer*, 44(9): 51–58, Sept 2011.

[27] O. Vermesan, M. Harrison, H. Vogt, K. Kalaboukas, M. Tomasella, K. Wouters, S. Gusmeroli, and S. Haller. *Vision and Challenges for Realising the Internet of Things*. European Commission, Information Society and Media, 2010.

[28] R. H. Weber. Internet of Things – New security and privacy challenges. *Computer Law & Security Review*, 26(1): 23–30, 2010.

[29] J. S. Winter. Surveillance in Ubiquitous Network Societies: Normative Conflicts Related to the Consumer In-store Supermarket Experience in the Context of the Internet of Things. *Ethics and Inf. Technol.*, 161: 27–41, 2014.

[30] A. Yao. How to generate and exchange secrets. In *Proc. of 27th FOCS, 1986.*, pages 162–167. IEEE, 1986.

[31] X. Zhu. Semi-Supervised Learning Literature Survey. http://pages.cs.wisc.edu/~jerryzhu/pub/ssl_survey.pdf, July 2008.

Biographies

T. Meskanen had his PhD in 2005. Since then he has been working as a researcher and lecturer at University of Turku. His main research interests are cryptography and public choice theory. His email address is tommes@utu.fi.

V. Niemi is a Professor of Mathematics at the University of Turku, Finland. Between 1997 and 2012 he was with Nokia Research Center in various positions, based in Finland and Switzerland. Niemi was also the chairman of the security standardization group of 3GPP during 2003–2009. His research interests include cryptography and mobile security. Valtteri can be contacted at valtteri.niemi@utu.fi.

N. Nieminen is a doctoral student at Turku Centre for Computer Science, Department of Mathematics and Statistics at the University of Turku. Her research interests include cryptography and its applications. Contact her at nmniem@utu.fi.

Cyber Security and the Internet of Things: Vulnerabilities, Threats, Intruders and Attacks

Mohamed Abomhara and Geir M. Køien

Department of Information and Communication Technology,
University of Agder, Norway
Corresponding Authors: {Mohamed.abomhara; geir.koien}@uia.no

Received 14 September 2014; Accepted 17 April 2015;
Publication 22 May 2015

Abstract

Internet of Things (IoT) devices are rapidly becoming ubiquitous while IoT services are becoming pervasive. Their success has not gone unnoticed and the number of threats and attacks against IoT devices and services are on the increase as well. Cyber-attacks are not new to IoT, but as IoT will be deeply interwoven in our lives and societies, it is becoming necessary to step up and take cyber defense seriously. Hence, there is a real need to secure IoT, which has consequently resulted in a need to comprehensively understand the threats and attacks on IoT infrastructure. This paper is an attempt to classify threat types, besides analyze and characterize intruders and attacks facing IoT devices and services.

Keywords: Internet of Things, Cyber-attack, Security threats.

1 Introduction

The recent rapid development of the Internet of Things (IoT) [1, 2] and its ability to offer different types of services have made it the fastest growing technology, with huge impact on social life and business environments. IoT has

Journal of Cyber Security, Vol. 4, 65–88.
doi: 10.13052/jcsm2245-1439.414

gradually permeated all aspects of modern human life, such as education, healthcare, and business, involving the storage of sensitive information about individuals and companies, financial data transactions, product development and marketing.

The vast diffusion of connected devices in the IoT has created enormous demand for robust security in response to the growing demand of millions or perhaps billions of connected devices and services worldwide [3–5].

The number of threats is rising daily, and attacks have been on the increase in both number and complexity. Not only is the number of potential attackers along with the size of networks growing, but the tools available to potential attackers are also becoming more sophisticated, efficient and effective [6, 7]. Therefore, for IoT to achieve fullest potential, it needs protection against threats and vulnerabilities [8].

Security has been defined as a process to protect an object against physical damage, unauthorized access, theft, or loss, by maintaining high confidentiality and integrity of information about the object and making information about that object available whenever needed [7, 9]. According to Kizza [7] there is no thing as the secure state of any object, tangible or not, because no such object can ever be in a perfectly secure state and still be useful. An object is secure if the process can maintain its maximum intrinsic value under different conditions. Security requirements in the IoT environment are not different from any other ICT systems. Therefore, ensuring IoT security requires maintaining the highest intrinsic value of both tangible objects (devices) and intangible ones (services, information and data).

This paper seeks to contribute to a better understanding of threats and their attributes (motivation and capabilities) originating from various intruders like organizations and intelligence. The process of identifying threats to systems and system vulnerabilities is necessary for specifying a robust, complete set of security requirements and also helps determine if the security solution is secure against malicious attacks [10]. As well as users, governments and IoT developers must ultimately understand the threats and have answers to the following questions:

1. What are the assets?
2. Who are the principal entities?
3. What are the threats?
4. Who are the threat actors?
5. What capability and resource levels do threat actors have?
6. Which threats can affect what assets?

7. Is the current design protected against threats?
8. What security mechanisms could be used against threats?

The remainder of this paper is organized as follows. Section 2 provides a background, definitions, and the primary security and privacy goals. Section 3 identifies some attacker motivations and capabilities, and provides an outline of various sorts of threat actors. Finally, the paper concludes with Section 4.

2 Background

The IoT [1, 2, 11] is an extension of the Internet into the physical world for interaction with physical entities from the surroundings. Entities, devices and services [12] are key concepts within the IoT domain, as depicted in Figure 1 [13]. They have different meanings and definitions among various projects. Therefore, it is necessary to have a good understanding of what IoT entities, devices and services are (discussed in detail in Section 2.1).

An entity in the IoT could be a human, animal, car, logistic chain item, electronic appliance or a closed or open environment [14]. Interaction among

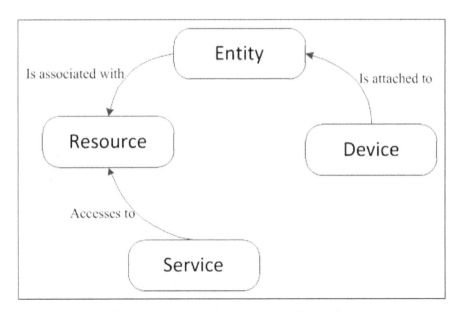

Figure 1 IoT model: key concepts and interactions.

entities is made possible by hardware components called devices [12] such as mobile phones, sensors, actuators or RFID tags, which allow the entities to connect to the digital world [15].

In the current state of technology, Machine-to-Machine (M2M) is the most popular application form of IoT. M2M is now widely employed in power, transportation, retail, public service management, health, water, oil and other industries to monitor and control the user, machinery and production processes in the global industry and so on [5, 16, 17]. According to estimates M2M applications will reach 12 billion connections by 2020 and generate approximately 714 billion euros in revenues [2].

Besides all the IoT application benefits, several security threats are observed [17–19]. The connected devices or machines are extremely valuable to cyber-attackers for several reasons:

1. Most IoT devices operate unattended by humans, thus it is easy for an attacker to physically gain access to them.
2. Most IoT components communicate over wireless networks where an attacker could obtain confidential information by eavesdropping.
3. Most IoT components cannot support complex security schemes due to low power and computing resource capabilities.

In addition, cyber threats could be launched against any IoT assets and facilities, potentially causing damage or disabling system operation, endangering the general populace or causing severe economic damage to owners and users [20, 21]. Examples include attacks on home automation systems and taking control of heating systems, air conditioning, lighting and physical security systems. The information collected from sensors embedded in heating or lighting systems could inform the intruder when somebody is at home or out. Among other things, cyber-attacks could be launched against any public infrastructure like utility systems (power systems or water treatment plants) [22] to stop water or electricity supply to inhabitants.

Security and privacy issues are a growing concern for users and suppliers in their shift towards the IoT [23]. It is certainly easy to imagine the amount of damage caused if any connected devices were attacked or corrupted. It is well-recognized that adopting any IoT technology within our homes, work, or business environments opens doors to new security problems. Users and suppliers must consider and be cautious with such security and privacy concerns.

2.1 Understanding IoT Devices and Services

In this section, the main IoT domain concepts that are important from a business process perspective are defined and classified, and the relationships between IoT components (IoT devices and IoT services) are described.

2.1.1 IoT device

This is a hardware component that allows the entity to be a part of the digital world [12]. It is also referred to as a smart thing, which can be a home appliance, healthcare device, vehicle, building, factory and almost anything networked and fitted with sensors providing information about the physical environment (e.g., temperature, humidity, presence detectors, and pollution), actuators (e.g., light switches, displays, motor-assisted shutters, or any other action that a device can perform) and embedded computers [24, 25].

An IoT device is capable of communicating with other IoT devices and ICT systems. These devices communicate via different means including cellular (3G or LTE), WLAN, wireless or other technologies [8]. IoT device classification depends on size, i.e., small or normal; mobility, i.e., mobile or fixed; external or internal power source; whether they are connected intermittently or always-on; automated or non-automated; logical or physical objects; and lastly, whether they are IP-enabled objects or non IP objects.

The characteristics of IoT devices are their ability to actuate and/or sense, the capability of limiting power/energy, connection to the physical world, intermittent connectivity and mobility [23]. Some must be fast and reliable and provide credible security and privacy, while others might not [9]. A number of these devices have physical protection whereas others are unattended.

In fact, in IoT environments, devices should be protected against any threats that can affect their functionality. However, most IoT devices are vulnerable to external and internal attacks due to their characteristics [16]. It is challenging to implement and use a strong security mechanism due to resource constraints in terms of IoT computational capabilities, memory, and battery power [26].

2.1.2 IoT services

IoT services facilitate the easy integration of IoT entities into the service-oriented architecture (SOA) world as well as service science [27]. According to Thoma [28], an IoT service is a transaction between two parties: the service provider and service consumer. It causes a prescribed function, enabling interaction with the physical world by measuring the state of entities or by initiating actions that will initiate a change to the entities.

A service provides a well-defined and standardized interface, offering all necessary functionalities for interacting with entities and related processes. The services expose the functionality of a device by accessing its hosted resources [12].

2.1.3 Security in IoT devices and services

Ensuring the security entails protecting both IoT devices and services from unauthorized access from within the devices and externally. Security should protect the services, hardware resources, information and data, both in transition and storage. In this section, we identified three key problems with IoT devices and services: data confidentiality, privacy and trust.

Data confidentiality represents a fundamental problem in IoT devices and services [27]. In IoT context not only user may access to data but also authorized object. This requires addressing two important aspects: first, access control and authorization mechanism and second authentication and identity management (IdM) mechanism. The IoT device needs to be able to verify that the entity (person or other device) is authorized to access the service. Authorization helps determine if upon identification, the person or device is permitted to receive a service. Access control entails controlling access to resources by granting or denying means using a wide array of criteria. Authorization and access control are important to establishing a secure connection between a number of devices and services. The main issue to be dealt with in this scenario is making access control rules easier to create, understand and manipulate. Another aspect that should be consider when dealing with confidentiality is authentication and identity management. In fact this issue is critical in IoT, because multiple users, object/things and devices need to authenticate each other through trustable services. The problem is to find solution for handling the identity of user, things/objects and devices in a secure manner.

Privacy is an important issue in IoT devices and service on account of the ubiquitous character of the IoT environment. Entities are connected, and data is communicated and exchanged over the internet, rendering user privacy a sensitive subject in many research works. Privacy in data collection, as well as data sharing and management, and data security matters remain open research issues to be fulfilled.

Trust plays an important role in establishing secure communication when a number of things communicate in an uncertain IoT environment. Two dimensions of trust should be considered in IoT: trust in the interactions between entities, and trust in the system from the users perspective [29] According to Køien [9] the trustworthiness of an IoT device depends on the device components including the hardware, such as processor, memory, sensors and actuators, software resources like hardware-based software, operating system, drivers and applications, and the power source. In order to gain user/services trust, there should be an effective mechanism of defining trust in a dynamic and collaborative IoT environment.

2.2 Security Threats, Attacks, and Vulnerabilities

Before addressing security threats, the system assets (system components) that make up the IoT must first be identified. It is important to understand the asset inventory, including all IoT components, devices and services.

An asset is an economic resource, something valuable and sensitive owned by an entity. The principal assets of any IoT system are the system hardware (include buildings, machinery, etc.) [11], software, services and data offered by the services [30].

2.2.1 Vulnerability

Vulnerabilities are weaknesses in a system or its design that allow an intruder to execute commands, access unauthorized data, and/or conduct denial-of-service attacks [31, 32]. Vulnerabilities can be found in variety of areas in the IoT systems. In particular, they can be weaknesses in system hardware or software, weaknesses in policies and procedures used in the systems and weaknesses of the system users themselves [7].

IoT systems are based on two main components; system hardware and system software, and both have design flaws quite often. Hardware vulnerabilities are very difficult to identify and also difficult to fix even if the vulnerability were identified due to hardware compatibility and interoperability and also the effort it take to be fixed. Software vulnerabilities can be found in operating systems, application software, and control software like communication protocols and devices drives. There are a number of factors that lead to software design flaws, including human factors and software complexity. Technical vulnerabilities usually happen due to human weaknesses. Results of not understanding the requirements comprise starting the project without a plan, poor communication between developers and users, a lack of resources, skills, and knowledge, and failing to manage and control the system [7].

2.2.2 Exposure

Exposure is a problem or mistake in the system configuration that allows an attacker to conduct information gathering activities. One of the most challenging issues in IoT is resiliency against exposure to physical attacks. In the most of IoT applications, devices may be left unattended and likely to be placed in location easily accessible to attackers. Such exposure raises the possibility that an attacker might capture the device, extract cryptographic secrets, modify their programming, or replace them with malicious device under the control of the attacker [33].

2.2.3 Threats

A threat is an action that takes advantage of security weaknesses in a system and has a negative impact on it [34]. Threats can originate from two primary sources: humans and nature [35, 36]. Natural threats, such as earthquakes, hurricanes, floods, and fire could cause severe damage to computer systems. Few safeguards can be implemented against natural disasters, and nobody can prevent them from happening. Disaster recovery plans like backup and contingency plans are the best approaches to secure systems against natural threats. Human threats are those caused by people, such as malicious threats consisting of internal [37] (someone has authorized access) or external threats [38] (individuals or organizations working outside the network) looking to harm and disrupt a system. Human threats are categorized into the following:

- Unstructured threats consisting of mostly inexperienced individuals who use easily available hacking tools.
- Structured threats as people know system vulnerabilities and can understand, develop and exploit codes and scripts. An example of a structured threat is Advanced Persistent Threats (APT) [39]. APT is a sophisticated network attack targeted at high-value information in business and government organizations, such as manufacturing, financial industries and national defense, to steal data [40].

As IoT become a reality, a growing number of ubiquitous devices has raise the number of the security threats with implication for the general public. Unfortunately, IoT comes with new set of security threat. There are a growing awareness that the new generation of smart-phone, computers and other devices could be targeted with malware and vulnerable to attack.

2.2.4 Attacks

Attacks are actions taken to harm a system or disrupt normal operations by exploiting vulnerabilities using various techniques and tools. Attackers launch attacks to achieve goals either for personal satisfaction or recompense. The measurement of the effort to be expended by an attacker, expressed in terms of their expertise, resources and motivation is called attack cost [32]. Attack actors are people who are a threat to the digital world [6]. They could be hackers, criminals, or even governments [7]. Additional details are discussed in Section 3.

An attack itself may come in many forms, including active network attacks to monitor unencrypted traffic in search of sensitive information; passive attacks such as monitoring unprotected network communications to decrypt weakly encrypted traffic and getting authentication information; close-in attacks; exploitation by insiders, and so on. Common cyber-attack types are:

(a) Physical attacks: This sort of attack tampers with hardware components. Due to the unattended and distributed nature of the IoT, most devices typically operate in outdoor environments, which are highly susceptible to physical attacks.

(b) Reconnaissance attacks – unauthorized discovery and mapping of systems, services, or vulnerabilities. Examples of reconnaissance attacks are scanning network ports [41], packet sniffers [42], traffic analysis, and sending queries about IP address information.

(c) Denial-of-service (DoS): This kind of attack is an attempt to make a machine or network resource unavailable to its intended users. Due to low memory capabilities and limited computation resources, the majority of devices in IoT are vulnerable to resource enervation attacks.

(d) Access attacks – unauthorized persons gain access to networks or devices to which they have no right to access. There are two different types of access attack: the first is physical access, whereby the intruder can gain access to a physical device. The second is remote access, which is done to IP-connected devices.

(e) Attacks on privacy: Privacy protection in IoT has become increasingly challenging due to large volumes of information easily available through remote access mechanisms. The most common attacks on user privacy are:

- Data mining: enables attackers to discover information that is not anticipated in certain databases.
- Cyber espionage: using cracking techniques and malicious software to spy or obtain secret information of individuals, organizations or the government.
- Eavesdropping: listening to a conversation between two parties [43].
- Tracking: a users movements can be tracked by the devices unique identification number (UID). Tracking a users location facilitates identifying them in situations in which they wish to remain anonymous.
- Password-based attacks: attempts are made by intruders to duplicate a valid user password. This attempt can be made in two different ways: 1) dictionary attack – trying possible combinations of letters and numbers to guess user passwords; 2) brute force attacks – using cracking tools to try all possible combinations of passwords to uncover valid passwords.

(f) Cyber-crimes: The Internet and smart objects are used to exploit users and data for materialistic gain, such as intellectual property theft, identity theft, brand theft, and fraud [6, 7, 44].

(g) Destructive attacks: Space is used to create large-scale disruption and destruction of life and property. Examples of destructive attacks are terrorism and revenge attacks.

(h) Supervisory Control and Data Acquisition (SCADA) Attacks: As any other TCP/IP systems, the SCADA [45] system is vulnerable to many cyber attacks [46, 47]. The system can be attacked in any of the following ways:

  i. Using denial-of-service to shut down the system.
  ii. Using Trojans or viruses to take control of the system. For instance, in 2008 an attack launched on an Iranian nuclear facility in Natanz using a virus named Stuxnet [48].

2.3 Primary Security and Privacy Goals

To succeed with the implementation of efficient IoT security, we must be aware of the primary security goals as follows:

2.3.1 Confidentiality

Confidentiality is an important security feature in IoT, but it may not be mandatory in some scenarios where data is presented publicly [18]. However, in most situations and scenarios sensitive data must not be disclosed or read by unauthorized entities. For instance patient data, private business data, and/or military data as well as security credentials and secret keys, must be hidden from unauthorized entities.

2.3.2 Integrity

To provide reliable services to IoT users, integrity is a mandatory security property in most cases. Different systems in IoT have various integrity requirements [49]. For instance, a remote patient monitoring system will have high integrity checking against random errors due to information sensitivities. Loss or manipulation of data may occur due to communication, potentially causing loss of human lives [6].

2.3.3 Authentication and authorization

Ubiquitous connectivity of the IoT aggravates the problem of authentication because of the nature of IoT environments, where possible communication would take place between device to device (M2M), human to device, and/or human to human. Different authentication requirements necessitate different solutions in different systems. Some solutions must be strong, for example authentication of bank cards or bank systems. On the other hand, most will have to be international, e.g., ePassport, while others have to be local [6]. The authorization property allows only authorized entities (any authenticated entity) to perform certain operations in the network.

2.3.4 Availability

A user of a device (or the device itself) must be capable of accessing services anytime, whenever needed. Different hardware and software components in IoT devices must be robust so as to provide services even in the presence of malicious entities or adverse situations. Various systems have different availability requirements. For instance, fire monitoring or healthcare monitoring systems would likely have higher availability requirements than roadside pollution sensors.

2.3.5 Accountability

When developing security techniques to be used in a secure network, accountability adds redundancy and responsibility of certain actions, duties and planning of the implementation of network security policies. Accountability itself cannot stop attacks but is helpful in ensuring the other security techniques are working properly. Core security issues like integrity and confidentiality may be useless if not subjected to accountability. Also, in case of a repudiation incident, an entity would be traced for its actions through an accountability process that could be useful for checking the inside story of what happened and who was actually responsible for the incident.

2.3.6 Auditing

A security audit is a systematic evaluation of the security of a device or service by measuring how well it conforms to a set of established criteria. Due to many bugs and vulnerabilities in most systems, security auditing plays an important role in determining any exploitable weaknesses that put the data at risk. In IoT, a systems need for auditing depends on the application and its value.

2.3.7 Non-repudiation

The property of non-repudiation produces certain evidence in cases where the user or device cannot deny an action. Non-repudiation is not considered an important security property for most of IoT. It may be applicable in certain contexts, for instance, payment systems where users or providers cannot deny a payment action.

2.3.8 Privacy goals

Privacy is an entitys right to determine the degree to which it will interact with its environment and to what extent the entity is willing to share information about itself with others. The main privacy goals in IoT are:

- Privacy in devices – depends on physical and commutation privacy. Sensitive information may be leaked out of the device in cases of device theft or loss and resilience to side channel attacks.
- Privacy during communication – depends on the availability of a device, and device integrity and reliability. IoT devices should communicate only when there is need, to derogate the disclosure of data privacy during communication.
- Privacy in storage – to protect the privacy of data stored in devices, the following two things should be considered:
  - Possible amounts of data needed should be stored in devices.

- Regulation must be extended to provide protection of user data after end-of-device life (deletion of the device data (Wipe) if the device is stolen, lost or not in use).
- Privacy in processing – depends on device and communication integrity [50]. Data should be disclosed to or retained from third parties without the knowledge of the data owner.
- Identity privacy – the identity of any device should only discovered by authorized entity (human/device).
- location privacy – the geographical position of relevant device should only discovered by authorized entity (human/device) [51].

3 Intruders, Motivations and Capabilities

Intruders have different motives and objectives, for instance, financial gain, influencing public opinion, and espionage, among many others. The motives and goals of intruders vary from individual attackers to sophisticated organized-crime organizations.

Intruders also have different levels of resources, skill, access and risk tolerance leading to the portability level of an attack occurring [52]. An insider has more access to a system than outsiders. Some intruders are well-funded and others work on a small budget or none. Every attacker chooses an attack that is affordable, an attack with good return on the investment based on budget, resources and experience [6]. In this section, intruders are categorized according to characteristics, motives and objectives, capabilities and resources.

3.1 Purpose and Motivation of Attack

Government websites, financial systems, news and media websites, military networks, as well as public infrastructure systems are the main targets for cyber-attacks. The value of these targets is difficult to estimate, and estimation often varies between attacker and defender. Attack motives range from identity theft, intellectual property theft, and financial fraud, to critical infrastructure attacks. It is quite difficult to list what motivates hackers to attack systems. For instance, stealing credit card information has become a hackers hobby nowadays, and electronic terrorism organizations attack government systems in order to make politics, religion interest.

3.2 Classification of Possible Intruders

A Dolev-Yao (DY) type of intruder shall generally be assumed [53, 54]. That is, an intruder which is in effect the network and which may intercept all or any message ever transmitted between IoT devices and hubs. The DY intruder is extremely capable but its capabilities are slightly unrealistic. Thus, safety will be much stronger if our IoT infrastructure is designed to be DY intruder resilient. However, the DY intruder lacks one capability that ordinary intruders may have, namely, physical compromise. Thus, tamper-proof devices are also greatly desirable. This goal is of course unattainable, but physical tamper resistance is nevertheless a very important goal, which, together with tamper detection capabilities (tamper evident) may be a sufficient first-line defense.

In the literature intruders are classified into two main types: internal and external. Internal intruders are users with privileges or authorized access to a system with either an account on a server or physical access to the network [21, 37]. External intruders are people who do not belong to the network domain. All intruders, whether internal or external, can be organized in many ways and involve individual attackers to spy agencies working for a country. The impact of an intrusion depends on the goals to be achieved. An individual attacker could have small objectives while spy agencies could have larger motives [55]. The various types of intruders will be discussed hereby based on their numbers, motives and objectives.

3.2.1 Individuals

Individual hackers are professionals who work alone and only target systems with low security [55]. They lack resources or expertise of professional hacking teams, organizations or spy agencies. Individual hacker targets are relatively small in size or diversity and the attacks launched have relatively lower impact than ones launched by organized groups (discussed in 3.2.2). Social engineering techniques are most commonly used by individual attackers, as they have to obtain basic information about a target system like the address, password, port information, etc. Public and social media websites are the most common places where general users can be deceived by hackers. Moreover, operating systems used on laptops, PCs, and mobile phones have common and known vulnerabilities exploitable by individual attackers.

Financial institutions such as banks are also major targets for individual attackers as they know that such types of networks carry financial transactions that can be hacked, and thus attackers can manipulate the information in their interest. Credit card information theft has a long history with individual hackers. With the growth of e-commerce, it is easier to use stolen credit card information to buy goods and services.

Individual hackers use tools such as viruses, worms and sniffers to exploit a system. They plan attacks based on equipment availability, internet access availability, the network environment and system security.

One of the individual hacker categories is the insider [21, 37]. Insiders are authorized individuals working against a system using insider knowledge or privileges. Insiders could provide critical information for outsider attackers (third party) to exploit vulnerabilities that can enable an attack. They know the weak points in the system and how the system works. Personal gain, revenge, and financial gain can motivate an insider. They can tolerate risk ranging from low to high depending on their motivation.

3.2.2 Organized groups

Criminal groups are becoming more familiar with ongoing communications and IoT technology. In addition, as they become more comfortable with technological applications, these groups can be more aware of opportunities offered by the infrastructure routing information of different networks. The motivations of these groups are quite diverse; their targets typically include particular organizations for revenge, theft of trade secrets, economic espionage, and targeting the national information infrastructure. They also involve selling personal information, such as financial data, to other criminal organizations, terrorists, and even governments.

They are very capable in terms of financial funding, expertise and resources. Criminal groups capabilities in terms of methods and techniques are moderate to high depending on what the goals are. They are very skillful at creating botnets and malicious software (e.g., computer viruses and scareware) and denial-of-service attack methods [44]. Organized criminals are likely to have access to funds, meaning they can hire skilled hackers if necessary, or purchase point-and-click attack tools from the underground economy with which to attack any systems [46]. Such criminals can tolerate higher risk than individual hackers and are willing to invest in profitable attacks.

Cyber terrorism [21, 56] is a form of cyber-attack that targets military systems, banks, and specific facilities such as satellites, and telecommunication systems associated with the national information infrastructure based on religious and political interests. Terrorist organizations depend on the internet to spread propaganda, raise funds, gather information, and communicate with co-conspirators in all parts of the world. Another prevalent group of criminal organization entails hacktivists. Hacktivists are groups of hackers who engage in activities such as denial-of-service, fraud, and/or identity theft. Also, some of these groups have political motivations, like the Syrian Electronic Army (SEA) [57], Iranian Cyber Army and Chinese cyber-warfare units [58].

3.2.3 Intelligence agency

Intelligence agencies from different countries are persistent in their efforts to probe the military systems of other countries for specific purposes, for example industrial espionage, and political and military espionage. To accomplish their objectives, the agencies require a large number of experts, infrastructure ranging from research and development entities to provide technologies and methodologies (hardware, software, and facilities) besides financial and human resources.

Such agencies have organized structures and sophisticated resources to accomplish their intrusion goals. This sort of agencies are the biggest threat to networks and necessitate tight surveillance and monitoring approaches to safeguard against threats to the information systems of prime importance for any country and military establishment.

4 Discussion and Conclusions

4.1 Discussion

The exponential growth of the IoT has led to greater security and privacy risks. Many such risks are attributable to device vulnerabilities that arise from cybercrime by hackers and improper use of system resources. The IoT needs to be built in such a way as to ensure easy and safe usage control. Consumers need confidence to fully embrace the IoT in order to enjoy its benefits and avoid security and privacy risks.

The majority of IoT devices and services are exposed to a number of common threats as discussed earlier, like viruses and denial-of-service attacks. Taking simple steps to avoid such threats and dealing with system vulnerabilities is not sufficient; thus, ensuring a smooth policy implementation process supported by strong procedures is needed.

The security development process requires thorough understanding of a systems assets, followed by identifying different vulnerabilities and threats that can exist. It is necessary to identify what the system assets are and what the assets should be protected against. In this paper, assets were defined as all valuable things in the system, tangible and intangible, which require protection. Some general, IoT assets include system hardware, software, data and information, as well as assets related to services, e.g. service reputation. It has been shown that it is crucial to comprehend the threats and system weaknesses in order to allocate better system mitigation. In addition, understanding potential attacks allows system developers to better determine where funds should be spent. Most commonly known threats have been described as DoS, physical attacks and attacks on privacy.

Three different types of intruders were discussed in this paper, namely individual attacks, organized groups, and intelligence agencies. Each attacker type has different skill levels, funding resources, motivation, and risk tolerance. It is very important to study the various types of attack actors and determine which are most likely to attack a system. Upon describing and documenting all threats and respective actors, it is easier to perceive which threat could exploit what weakness in the system. Generally, it is assumed that IoT intruder has full DY intruder capabilities in addition to some limited physical compromise power. We will presume that physical compromise attacks do not scale, and they will therefore only at-worst affect a limited population of the total number of IoT devices. IoT architecture must consequently be designed to cope with compromised devices and be competent in detecting such incidents. It is concluded that attackers employ various methods, tools, and techniques to exploit vulnerabilities in a system to achieve their goals or objectives. Understanding attackers motives and capabilities is important for an organization to prevent potential damage. To reduce both potential threats and their consequences, more research is needed to fill the gaps in knowledge regarding threats and cybercrime and provide the necessary steps to mitigate probable attacks.

5 Conclusions

IoT faces a number of threats that must be recognized for protective action to be taken. In this paper, security challenges and security threats to IoT were introduced. The overall goal was to identify assets and document potential threats, attacks and vulnerabilities faced by the IoT.

An overview of the most important IoT security problems was provided, with particular focus on security challenges surrounding IoT devices and services. Security challenges, such as confidentiality, privacy and entity trust were identified. We showed that in order to establish more secure and readily available IoT devices and services, security and privacy challenges need to be addressed. The discussion also focused upon the cyber threats comprising actors, motivation, and capability fuelled by the unique characteristics of cyberspace. It was demonstrated that threats from intelligence agencies and criminal groups are likely to be more difficult to defeat than those from individual hackers. The reason is that their targets may be much less predictable while the impact of an individual attack is expected to be less severe.

It was concluded that much work remains to be done in the area of IoT security, by both vendors and end-users. It is important for upcoming standards to address the shortcomings of current IoT security mechanisms. As future work, the aim is to gain deeper understanding of the threats facing IoT infrastructure as well as identify the likelihood and consequences of threats against IoT. Definitions of suitable security mechanisms for access control, authentication, identity management, and a flexible trust management framework should be considered early in product development. We hope this survey will be useful to researchers in the security field by helping identify the major issues in IoT security and providing better understanding of the threats and their attributes originating from various intruders like organizations and intelligence agencies.

References

[1] L. Atzori, A. Iera, and G. Morabito, "The internet of things: A survey," *Computer networks*, vol. 54, no. 15, pp. 2787–2805, 2010.

[2] S. Andreev and Y. Koucheryavy, "Internet of things, smart spaces, and next generation networking," *Springer, LNCS*, vol. 7469, p. 464, 2012.

[3] J. S. Kumar and D. R. Patel, "A survey on internet of things: Security and privacy issues," *International Journal of Computer Applications*, vol. 90, no. 11, pp. 20–26, March 2014, published by Foundation of Computer Science, New York, USA.

[4] A. Stango, N. R. Prasad, and D. M. Kyriazanos, "A threat analysis methodology for security evaluation and enhancement planning," in *Emerging Security Information, Systems and Technologies, 2009. SECURWARE'09. Third International Conference on*. IEEE, 2009, pp. 262–267.

[5] D. Jiang and C. ShiWei, "A study of information security for m2m of iot," in *Advanced Computer Theory and Engineering (ICACTE), 2010 3rd International Conference on*, vol. 3. IEEE, 2010, pp. V3–576.

[6] B. Schneier, *Secrets and lies: digital security in a networked world*. John Wiley & Sons, 2011.

[7] J. M. Kizza, *Guide to Computer Network Security*. Springer, 2013.

[8] M. Taneja, "An analytics framework to detect compromised iot devices using mobility behavior," in *ICT Convergence (ICTC), 2013 International Conference on*. IEEE, 2013, pp. 38–43.

[9] G. M. Koien and V. A. Oleshchuk, *Aspects of Personal Privacy in Communications-Problems, Technology and Solutions*. River Publishers, 2013.

[10] N. R. Prasad, "Threat model framework and methodology for personal networks (pns)," in *Communication Systems Software and Middleware, 2007. COMSWARE 2007. 2nd International Conference on*. IEEE, 2007, pp. 1–6.

[11] O. Vermesan, P. Friess, P. Guillemin, S. Gusmeroli, H. Sundmaeker, A. Bassi, I. S. Jubert, M. Mazura, M. Harrison, M. Eisenhauer *et al.* "Internet of things strategic research roadmap," *Internet of Things-Global Technological and Societal Trends*, pp. 9–52, 2011.

[12] S. De, P. Barnaghi, M. Bauer, and S. Meissner, "Service modelling for the internet of things," in *Computer Science and Information Systems (FedCSIS), 2011 Federated Conference on*. IEEE, 2011, pp. 949–955.

[13] G. Xiao, J. Guo, L. Xu, and Z. Gong, "User interoperability with heterogeneous iot devices through transformation," 2014.

[14] J. Gubbi, R. Buyya, S. Marusic, and M. Palaniswami, "Internet of things (iot): A vision, architectural elements, and future directions," *Future Generation Computer Systems*, vol. 29, no. 7, pp. 1645–1660, 2013.

[15] M. Zorzi, A. Gluhak, S. Lange, and A. Bassi, "From today's intranet of things to a future internet of things: a wireless-and mobility-related view," *Wireless Communications, IEEE*, vol. 17, no. 6, pp. 44–51, 2010.

[16] C. Hongsong, F. Zhongchuan, and Z. Dongyan, "Security and trust research in m2m system," in *Vehicular Electronics and Safety (ICVES), 2011 IEEE International Conference on*. IEEE, 2011, pp. 286–290.

[17] I. Cha, Y. Shah, A. U. Schmidt, A. Leicher, and M. V. Meyerstein, "Trust in m2m communication," *Vehicular Technology Magazine, IEEE*, vol. 4, no. 3, pp. 69–75, 2009.

[18] J. Lopez, R. Roman, and C. Alcaraz, "Analysis of security threats, requirements, technologies and standards in wireless sensor networks,"

in *Foundations of Security Analysis and Design V.* Springer, 2009, pp. 289–338.

[19] R. Roman, J. Zhou, and J. Lopez, "On the features and challenges of security and privacy in distributed internet of things," *Computer Networks*, vol. 57, no. 10, pp. 2266–2279, 2013.

[20] Y. Cheng, M. Naslund, G. Selander, and E. Fogelstrom, "Privacy in machine-to-machine communications a state-of-the-art survey," in *Communication Systems (ICCS), 2012 IEEE International Conference on.* IEEE, 2012, pp. 75–79.

[21] M. Rudner, "Cyber-threats to critical national infrastructure: An intelligence challenge," *International Journal of Intelligence and CounterIntelligence*, vol. 26, no. 3, pp. 453–481, 2013.

[22] R. Kozik and M. Choras, "Current cyber security threats and challenges in critical infrastructures protection," in *Informatics and Applications (ICIA), 2013 Second International Conference on.* IEEE, 2013, pp. 93–97.

[23] P. N. Mahalle, N. R. Prasad, and R. Prasad, "Object classification based context management for identity management in internet of things," *International Journal of Computer Applications*, vol. 63, no. 12, pp. 1–6, 2013.

[24] A. Gluhak, S. Krco, M. Nati, D. Pfisterer, N. Mitton, and T. Razafindralambo, "A survey on facilities for experimental internet of things research," *Communications Magazine, IEEE*, vol. 49, no. 11, pp. 58–67, 2011.

[25] Y. Benazzouz, C. Munilla, O. Gunalp, M. Gallissot, and L. Gurgen, "Sharing user iot devices in the cloud," in *Internet of Things (WF-IoT), 2014 IEEE World Forum on.* IEEE, 2014, pp. 373–374.

[26] G. M. Køien, "Reflections on trust in devices: an informal survey of human trust in an internet-of-things context," *Wireless Personal Communications*, vol. 61, no. 3, pp. 495–510, 2011.

[27] D. Miorandi, S. Sicari, F. De Pellegrini, and I. Chlamtac, "Internet of things: Vision, applications and research challenges," *Ad Hoc Networks*, vol. 10, no. 7, pp. 1497–1516, 2012.

[28] M. Thoma, S. Meyer, K. Sperner, S. Meissner, and T. Braun, "On iot-services: Survey, classification and enterprise integration," in *Green Computing and Communications (GreenCom), 2012 IEEE International Conference on.* IEEE, 2012, pp. 257–260.

[29] M. Abomhara and G. Koien, "Security and privacy in the internet of things: Current status and open issues," in *PRISMS 2014 The 2nd*

International Conference on Privacy and Security in Mobile Systems
(PRISMS 2014), Aalborg, Denmark, May 2014.

[30] D. Watts, "Security and vulnerability in electric power systems," in *35th North American power symposium*, vol. 2, 2003, pp. 559–566.

[31] D. L. Pipkin, *Information security*. Prentice Hall PTR, 2000.

[32] E. Bertino, L. D. Martino, F. Paci, and A. C. Squicciarini, "Web services threats, vulnerabilities, and countermeasures," in *Security for Web Services and Service-Oriented Architectures*. Springer, 2010, pp. 25–44.

[33] D. G. Padmavathi, M. Shanmugapriya *et al.*, "A survey of attacks, security mechanisms and challenges in wireless sensor networks," *arXiv preprint arXiv:0909.0576*, 2009.

[34] H. G. Brauch, "Concepts of security threats, challenges, vulnerabilities and risks," in *Coping with Global Environmental Change, Disasters and Security*. Springer, 2011, pp. 61–106.

[35] K. Dahbur, B. Mohammad, and A. B. Tarakji, "A survey of risks, threats and vulnerabilities in cloud computing," in *Proceedings of the 2011 International conference on intelligent semantic Web-services and applications*. ACM, 2011, p. 12.

[36] R. K. Rainer and C. G. Cegielski, *Introduction to information systems: Enabling and transforming business*. John Wiley & Sons, 2010.

[37] A. J. Duncan, S. Creese, and M. Goldsmith, "Insider attacks in cloud computing," in *Trust, Security and Privacy in Computing and Communications (TrustCom), 2012 IEEE 11th International Conference on*. IEEE, 2012, pp. 857–862.

[38] P. Baybutt, "Assessing risks from threats to process plants: Threat and vulnerability analysis," *Process Safety Progress*, vol. 21, no. 4, pp. 269–275, 2002.

[39] C. Tankard, "Advanced persistent threats and how to monitor and deter them," *Network security*, vol. 2011, no. 8, pp. 16–19, 2011.

[40] F. Li, A. Lai, and D. Ddl, "Evidence of advanced persistent threat: A case study of malware for political espionage," in *Malicious and Unwanted Software (MALWARE), 2011 6th International Conference on*. IEEE, 2011, pp. 102–109.

[41] S. Ansari, S. Rajeev, and H. Chandrashekar, "Packet sniffing: a brief introduction," *Potentials, IEEE*, vol. 21, no. 5, pp. 17–19, 2002.

[42] M. De Vivo, E. Carrasco, G. Isern, and G. O. de Vivo, "A review of port scanning techniques," *ACM SIGCOMM Computer Communication Review*, vol. 29, no. 2, pp. 41–48, 1999.

[43] I. Naumann and G. Hogben, "Privacy features of european eid card specifications," *Network Security*, vol. 2008, no. 8, pp. 9–13, 2008.

[44] C. Wilson, "Botnets, cybercrime, and cyberterrorism: Vulnerabilities and policy issues for congress." DTIC Document, 2008.

[45] A. Daneels and W. Salter, "What is scada," in *International Conference on Accelerator and Large Experimental Physics Control Systems*, 1999, pp. 339–343.

[46] A. Nicholson, S. Webber, S. Dyer, T. Patel, and H. Janicke, "Scada security in the light of cyber-warfare," *Computers & Security*, vol. 31, no. 4, pp. 418–436, 2012.

[47] V. M. Igure, S. A. Laughter, and R. D. Williams, "Security issues in scada networks," *Computers & Security*, vol. 25, no. 7, pp. 498–506, 2006.

[48] M. Kelleye, "Business Insider. The Stuxnet attack on Irans Nuclear Plant was Far more Dangerous Than Previously Thought," http://www.businessinsider.com/stuxnet-was-far-more-dangerous-than-previous-thought-2013-11/,2013, [Online; accessed 03-Sep-2014].

[49] B. Jung, I. Han, and S. Lee, "Security threats to internet: a korean multi-industry investigation," *Information & Management*, vol. 38, no. 8, pp. 487–498, 2001.

[50] C. P. Mayer, "Security and privacy challenges in the internet of things," *Electronic Communications of the EASST*, vol. 17, 2009.

[51] A. R. Beresford, "Location privacy in ubiquitous computing," *Computer Laboratory, University of Cambridge, Tech. Rep*, vol. 612, 2005.

[52] S. Pramanik, "Threat motivation," in *Emerging Technologies for a Smarter World (CEWIT), 2013 10th International Conference and Expo on*. IEEE, 2013, pp. 1–5.

[53] D. Dolev and A. C. Yao, "On the security of public key protocols," *Information Theory, IEEE Transactions on*, vol. 29, no. 2, pp. 198–208, 1983.

[54] I. Cervesato, "The dolev-yao intruder is the most powerful attacker," in *16th Annual Symposium on Logic in Computer ScienceLICS*, vol. 1. Citeseer, 2001.

[55] J. Sheldon, "State of the art: Attackers and targets in cyberspace," *Journal of Military and Strategic Studies*, vol. 14, no. 2, 2012.

[56] E. M. Archer, "Crossing the rubicon: Understanding cyber terrorism in the european context," *The European Legacy*, no. ahead-of-print, pp. 1–16, 2014.

[57] A. K. Al-Rawi, "Cyber warriors in the middle east: The case of the syrian electronic army," *Public Relations Review*, 2014.

[58] D. Ball, "Chinas cyber warfare capabilities," *Security Challenges*, vol. 7, no. 2, pp. 81–103, 2011.

Biographies

M. Abomhara is currently pursuing his PhD at University of Agder, Norway. His research work is in the area of computer security, information security, information system management, cyber-security, and Internet of things. He received a Master of Computer Science (Data Communication and Computer Network) from University of Malaya, Malaysia in 2011. He also received a Master of Business Administration (MBA, Information technology management) from Multimedia University, Malaysia in 2013 and a Bachelor of Computer Science from 7th October University, Libya in 2006.

G. M. Køien is an associate professor in security and privacy in ICT at the University of Agder, Norge. He has previously worked for Ericsson Norway, System Sikkerhet AS and Telenor R & D. During his time with Telenor R & D he was the Telenor delegate to the SA3 (3GPP) work group on security. He received his PhD for Aalborg University, Denmark in 2008.

www.ingramcontent.com/pod-product-compliance
Lightning Source LLC
LaVergne TN
LVHW012333060326
832902LV00011B/1862